SpringerBriefs in Latin American Studies

SpringerBriefs in Latin American Studies promotes quality scientific research focusing on Latin American countries. The series accepts disciplinary and interdisciplinary titles related to geographical, environmental, cultural, economic, political, urban and health-related research dedicated to Latin America. The Series will publish compact volumes (50 to 125 pages) refereed by a region or country expert specialized in Latin American studies. We offer fast publication time of 8 to 12 weeks after acceptance. The series aims to raise the profile of Latin American studies, showcasing important works developed focusing on the region. It is aimed at researchers, students, and everyone interested in Latin American topics.

Elisa Belfiori · Gastón Gertner
Editors

Climate Change and Economic Growth

Lessons from Argentina

Springer

Editors
Elisa Belfiori
School of Business
Universidad Torcuato Di Tella
Buenos Aires, Argentina

Gastón Gertner
Florence School of Transnational Governance
European University Institute
Florence, Italy

ISSN 2366-763X ISSN 2366-7648 (electronic)
SpringerBriefs in Latin American Studies
ISBN 978-3-031-99636-8 ISBN 978-3-031-99637-5 (eBook)
https://doi.org/10.1007/978-3-031-99637-5

This Springer imprint is published by the registered company Springer Nature Switzerland AG
The registered company address is: Gewerbestrasse 11, 6330 Cham, Switzerland

Foreword by Hernando Zuleta

Since the Industrial Revolution, economic growth has been both a promise and a driving force. It has helped reduce poverty, expand opportunities, and increase the productive capacity of nations. At the same time, it has raised a persistent concern: the possibility that the physical limits of the planet could halt progress. This tension between growth and natural constraints is not new. In *An Essay on the Principle of Population* (Malthus 1798), Thomas Malthus argued that limited resources would prevent prosperity from lasting. For more than a century, the expansion of industry and technology seemed to disprove this view. Yet in the second half of the twentieth century, concerns about limits to growth resurfaced, spurred by the energy crisis, the *Limits to Growth* report by the Club of Rome (Meadows et al. 1972), and growing evidence of global environmental degradation.

In the early twenty-first century, this tension has become more acute. Latin America and other developing regions need growth to reduce poverty, but growth that follows traditional patterns of energy use and resource extraction risks intensifying the climate crisis. The dilemma is clear: how can economic opportunities expand without undermining the natural foundations on which they depend?

Economists have long grappled with this question. Jevons (1865), Sidgwick (1887), Marshall (1890), and Pigou (1920) each considered the relationship between economic activity and natural limits. Pigou's idea of externalities was particularly influential in showing that markets alone do not capture the environmental costs of human actions. Later, Georgescu-Roegen (1971) emphasized the biophysical limits to growth. More recently, Nordhaus (1993, 1994, 1997) and Nordhaus and Boyer (1999) developed models that formally linked economic growth and climate change. Together, these contributions underline that GDP growth is insufficient as a sole metric: we must also consider how resources are used, how benefits are distributed, and how economic choices affect the climate system.

Latin America faces this challenge within a context of weak productivity, limited diversification, and persistent dependence on natural resource exports. Growth spurts tied to commodity booms have often been followed by stagnation. The region's unfinished growth agenda—achieving stable, long-term development—can no longer be separated from climate change. Although Latin America contributes a relatively small share of global emissions, it is highly vulnerable to their consequences: stronger hurricanes, prolonged droughts, shrinking glaciers, wildfires, and floods. Pursuing growth without sustainability risks worsening these threats.

The climate challenge also exposes a contradiction at the heart of development strategies. Latin America is rich in natural resources such as oil, gas, lithium, copper, and fertile agricultural land—resources that could fuel another phase of growth. Yet intensive use of these assets also drives emissions and environmental damage. Agriculture and livestock illustrate the tension clearly: they are central to the economy but also major contributors to greenhouse gas emissions. Expanding production without transforming practices would only increase the region's role in global warming.

In this context, the book *Climate Change and Economic Growth: Lessons from Argentina*, edited by Elisa Belfiori and Gastón Gertner, is both valuable and timely. It presents the region's environmental challenges with rigor and a clear orientation toward informing policy. Written in accessible language, it brings together diverse perspectives and highlights the contributions of young researchers. It is a work that combines academic seriousness with clarity of exposition and places on the agenda issues that can no longer be postponed.

The dilemma of growth in Latin America in the face of climate change does not have simple solutions. Addressing it requires redefining the role of the state and the private sector, reforming fiscal systems, promoting innovation, and strengthening regional cooperation. Growth and sustainability should not be seen as opposing goals, but as interdependent ones. This book contributes to that understanding. By offering diagnoses, policy proposals, and interdisciplinary perspectives, it encourages us to think about a new social contract for the region—one that links growth with sustainability, equity with efficiency, and innovation with intergenerational responsibility. In this sense, it is a call for more ambitious policies and a reminder that Latin America's future depends on how it confronts today's climate challenge.

August 2025

Hernando Zuleta
Professor and Dean, Department of Economics
Universidad de los Andes
Bogotá, Colombia

References

Georgescu-Roegen N (1971) The entropy law and the economic process. Harvard University Press

Jevons WS (1865) The coal question. Macmillan, London

Marshall A (1890) Principles of economics. Macmillan, London

Malthus TR (1798) An essay on the principle of population. J. Johnson, London

Meadows DH, Meadows DL, Randers J, Behrens WW (1972) The limits to growth. Universe Books, New York

Nordhaus WD (1993, 1994, 1997) Reflections on the economics of climate change and DICE and RICE models. Yale University
Nordhaus WD, Boyer J (1999) Warming the world: economic models of global warming. MIT Press
Pigou AC (1920) The economics of welfare. Macmillan, London
Sidgwick H (1887) The principles of political economy. Macmillan, London

Foreword by Massimo Filippini

The current energy systems of all countries are dominated by fossil fuels, which are the source of two fundamental problems for our societies, namely climate change and air pollution. To combat these critical problems, we need in-depth scientific studies that are capable of analyzing the consequences of these environmental problems and identifying and suggesting the best energy and climate policies. This interesting book, edited by Elisa Belfiori and Gastón Gertner, fits into this context, with rich multidisciplinary scientific contributions on topics related in particular to climate change and the Argentina situation.

In this foreword, I will highlight three key aspects of the economic discussion regarding policies implemented to address the challenges posed by climate change. The first aspect concerns the relationship between energy policy and climate policy. The second concerns the role of behavioral anomalies in the definition of energy and climate policies. The third aspect pertains to the lack of attention given in the economic literature on climate change to the benefits that climate and energy policies provide for air quality and energy security.

To reduce adverse effects on the environment and the economic and non-economic well-being, we must achieve the energy transition by implementing coordinated energy and climate policy measures. These two policies are often discussed, designed, and implemented in isolation. This separation is problematic because, as both domains share only a single overarching objective—reducing greenhouse gas emissions—other goals may diverge, creating tensions and inefficiencies. Energy policy has three main goals: energy security, affordability, and environmental sustainability. In contrast, climate policy focuses on reducing greenhouse gas emissions and promoting adaptation measures. The two policies share a common goal of environmental sustainability, which includes the reduction of greenhouse gases. However, their differing objectives can lead to conflicts, misaligned incentives, and costly coordination failures. For example, policymakers may choose to subsidize fossil fuels for the sake of affordability—something that has occurred recently in Europe due to the surge in gasoline and gas prices caused by the Russia-Ukraine war—while simultaneously implementing a carbon tax.

Recognizing these connections is essential because the energy transition—the shift from fossil-fuel-based systems to systems based on renewable energy sources and high energy efficiency—is not just an issue of energy policy or solely a climate policy initiative. It is one of the fundamental components in the fight against climate change and must be approached with both perspectives in mind. Harmonizing the design and implementation of energy and climate policy measures, which are often defined in separate legal instruments, can lower policy costs, enhance political acceptability, and speed up progress toward decarbonization and energy system resilience.

A second crucial dimension in policy design is often underemphasized: the role of behavioral anomalies. In a recent open-access textbook in energy economics and policy (Filippini and Srinivasan 2024), we propose to include behavioral anomalies among market participants as a market failure in addition to the classical market failures such as negative externalities and common resources. These anomalies refer to situations in which individuals or companies make economic decisions that diverge from the assumptions of neoclassical microeconomic theory, which posits rational behavior. The literature on behavioral economics indicates that individuals or firms, in their economic decisions, may demonstrate bounded rationality, meaning they make choices that are not optimal from a rational perspective. This bounded rationality behavior is due to several factors, such as limited cognitive ability to process all information or to perform optimization processes, or the presence of incomplete information.

The presence of boundedly rational individuals and firms can pose a significant barrier to achieving the energy transition. This transition requires substantial investments in renewable technologies and energy efficiency. If individuals do not act rationally and fail to evaluate the costs and benefits of a technology throughout its entire lifecycle, there is a serious risk that, without government intervention through standards or regulatory measures, they will not make the necessary investments. For example, even when there is a subsidy that makes adopting new energy-efficient technology financially appealing, households may still hesitate to adopt these technologies due to factors such as inertia, status quo bias, or a lack of awareness regarding the long-term benefits of energy efficiency. This phenomenon is referred to as the "energy efficiency gap." Similarly, a focus on short-term financial returns in investment decisions can hinder the adoption of low-carbon technologies that involve higher upfront costs but offer long-term savings through reduced energy expenses.

From an energy and climate policy point of view, the presence of behavioral anomalies has two significant implications. First, in the design of energy and climate policy instruments, it is essential to include targeted behavioral interventions—such as default rules, information framing, and social norm messaging. Additionally, we must be cautious about overreliance on market-oriented instruments such as carbon taxes or subsidies, which assume markets animated by rational economic agents. While such instruments remain essential, their effectiveness may be significantly reduced if market participants fail to respond to price signals as theory predicts. In summary, a contemporary energy and climate policy should combine market-based strategies with behavioral and regulatory approaches to effectively address real-world decision-making processes.

The final aspect of energy and climate policy that I want to discuss is the global and local benefits of the energy transition. The global benefits relate to reducing the adverse effects of climate change in all regions of the world, with some regions benefiting more than others. These benefits are primarily medium to long terms, making them less noticeable at the individual and regional level. The regional benefits include improving air quality, particularly in urban areas where a significant portion of the world's population lives, and enhancing energy security at the national level. These advantages are primarily local and national, offering immediate and tangible results that can be seen at both the individual and community levels.

In the scientific literature on climate and energy economics and policy, it appears that the narrative surrounding research problems and results tends to emphasize global benefits over local ones. In many studies, the improvement of air quality and energy security due to climate and energy policy is referred to as a "co-benefit," i.e., a secondary benefit. Further, numerous studies that evaluate the impact of climate policies on welfare often fail to quantify these significant regional benefits.

Implementing climate policies, such as taxes or bans, is often challenging due to low public acceptance. This lack of motivation could stem from the perception that global benefits are distant in both time and space. A key question in this context is whether emphasizing the local and national benefits of the energy transition—such as improved air quality and enhanced energy security—could increase public acceptance of these policies. The message could be framed as follows: "The energy transition enhances the quality of life in our country by improving air quality and increasing energy security, while also helping to mitigate the adverse effects of climate change." I believe that the role of narrative in the acceptance of climate and energy policy instruments is essential and, therefore, needs more attention from the scientific community.

August 2025

Prof. Dr. Massimo Filippini
ETH Zurich and Università della Svizzera Italiana, Switzerland
Director Centre for Energy Policy and Economics (CEPE), ETH Zurich
Zurich, Switzerland

Reference

Filippini M, Srinivasan S (2024) An introduction to energy economics and policy. Cambridge University Press

Prologue

Climate change is no longer a distant concern; it is a defining issue of our time, dominating discussions among policymakers, journalists, students, business leaders, and scientists alike. This urgency resonates across both developed and developing nations, underscoring the pressing need for equitable solutions that mitigate the burden on future generations. One of the many challenges that the climate problem brings is its broad, interdisciplinary dimension. One-sided methods often fall short, lacking the comprehensive evidence, support, and consensus required to confront climate change effectively. By integrating diverse perspectives, we can define, correct, and mitigate the environmental challenges our world faces—now and for the future.

In this book, we bring together researchers in multiple fields—from economics and political science to neuroscience and psychology—to contribute to an informed and research-based dialogue on climate change and sustainable growth. The book explores the challenges and opportunities involved in achieving an energy transition that decouples economic growth from the drivers of climate change. All contributors share a global south viewpoint in general and a particular focus on Argentina in their analyses.

The book is organized into three parts. Part I offers a global overview of the climate crisis and key policy responses. Chapter 1 examines the current state of international climate negotiations, highlighting the complex and evolving geopolitical landscape. Chapter 2 focuses on carbon markets, with special attention to pricing mechanisms in agriculture and livestock—sectors that are especially relevant to Argentina and the Latin American region.

Part II turns to the political economy of the energy transition. Chapter 3 provides a global perspective on the systemic shifts required for sustainable energy, while Chap. 4 zooms in on Latin America, revealing how political and economic constraints can impede progress, even in countries with significant renewable energy potential.

Part III explores how behavioral insights can be leveraged to bridge the gap between climate awareness and climate action. Chapters 5–7 examine the role of individual and collective behavior in shaping environmental outcomes in Argentina. Topics include behaviorally informed climate communication, energy use in informal urban settlements, and the growing influence of youth movements and social media in promoting environmental change. Together, these chapters offer practical strategies and comparative insights for policy design and public engagement.

Argentina faces a particularly complex challenge in addressing climate change, especially given its significant emissions from agriculture and livestock, and its ongoing macroeconomic struggles. The key question is how to foster sustainable economic growth while reducing inequality and protecting the environment. Meeting this challenge requires stronger collaboration between academic institutions and policymakers to develop innovative strategies, regulatory frameworks, and investments that advance sustainable development in Argentina and across Latin America.

Yet a noticeable gap remains in Latin America's—and Argentina's—academic discourse on climate change and related fields. This book aims to help close that gap. By offering a "safe space" for interdisciplinary collaboration, this volume aims to bridge the gap between research and policy and to support the development of more effective public policies. The book also seeks to stimulate conversations on regulatory frameworks and communications strategies that can facilitate a viable energy transition. In featuring Argentina as a central case study, the book provides a distinct perspective on climate change in Latin America. We hope that the research-based insights presented in these essays enrich academic resources and inspire broader action from governments, the private sector, and civil society. Our collective goal is to advance a more sustainable future for communities everywhere.

Buenos Aires, Argentina

Elisa Belfiori
Gastón Gertner

Acknowledgements This book, *Climate Change and Economic Growth: Lessons from Argentina*, is the result of the presentations and discussions that emerged during the conference Energy, Climate Change, and the Path to Sustainable Growth, held at Universidad Torcuato Di Tella in Buenos Aires in August 2022.

About This Book

This book examines the complex relationship between energy use, economic growth, and the environment, focusing on Argentina and its relevance to other Latin American countries and developing economies. It offers an interdisciplinary approach and policy recommendations to reconcile efforts to reduce, adapt, and mitigate climate change with sustainable development goals. Its interdisciplinary approach includes two chapters that lay on the intersection between behavioral science and climate change, highlighting the use of behavioral science to understand society's actions concerning climate change and how to reverse biases and promote more sustainable behaviors. The book also includes a unique contribution from young scholars, reflecting the growing demand of younger generations for intergenerational justice and concrete actions to protect the environment. With a broad audience in mind, including academics, researchers, policymakers, and the general public interested in climate change, energy, and sustainable development, this volume offers a distinctive perspective not found in other publications. Keywords: Argentina, climate change, developing economies, economic growth, energy transition, energy use, environment, interdisciplinary approach, intergenerational justice, Latin American countries, policy recommendations, sustainable development goals, Behavioral science, biases.

Contents

Part I
Climate Change: Global Overview and Carbon Markets

Chapter 1
Understanding the Current State of Climate Change Global Negotiations

Hernán Carlino and Micaela Carlino

Abstract The chapter is dedicated to an appraisal of the current state of climate change global negotiations. Such an appraisal requires understanding the evolution of the negotiation process, since its formal inception in 1992 during the Rio Summit, as well as considering the multifaceted implications of the efforts involved in the construction of climate change governance. Further, the international context in which negotiations are being carried out has undergone radical mutations in the last three decades, which have deeply modified the conditions in which the negotiation process is unfolding. The successive efforts to establish a climate regime with a legal form capable of navigating the increasing geo-economic clashes and still deliver a framework that would drive transformations in key global systems (i.e., energy, transport, industry, and food), aimed at collectively bending the curve of GHG emissions, demonstrate the political will to work cooperatively despite the intractable quest for dominance and unchallenged power in a world in heightened conflict. Presently, discourses of delay to justify inaction or legitimize scant efforts by nations, economic agents, and even societies, loom again because emissions are still rising after three decades of drawn-out negotiations while the political landscape is even more complex. However, the rate of growth of emissions has recently slowed. When working towards a positive narrative to accelerate action, it is worth acknowledging that efforts to mitigate climate change have been able to diminish emissions if compared to a world without climate policies and that a fair, robust, and effective climate regime is essential to challenge the status quo and overcome climate change.

H. Carlino (✉) · M. Carlino
Fundación Torcuato Di Tella, Buenos Aires, Argentina
e-mail: hcarl@itdt.edu

M. Carlino
e-mail: mc@itdt.edu

E. Belfiori and G. Gertner (eds.), *Climate Change and Economic Growth*,
SpringerBriefs in Latin American Studies, https://doi.org/10.1007/978-3-031-99637-5_1

1.1 Introduction

The contrast between commemorating in 2023 three decades of negotiations towards a United Nations climate change regime,[1] and almost simultaneously confirming that anthropogenic greenhouse gas emissions (GHG) still evince an upward trajectory, has posed numerous questions on the actual progress in negotiations and hesitations on the relevance of the outcomes achieved. Those doubts deepened despite acknowledging the divisive political, social, and economic context in which the negotiations process is unfolding.

While the world has gone through considerable changes since the inception of the United Nations Framework Convention on Climate Change (UNFCCC), the mutations in the regime seem not to have been able to entirely reflect the varying circumstances and the evolving needs of a rapidly changing world, even if vigorously attaining to do so.

This chapter aims to understand climate regime accomplishments, ponder progress achieved, and, in particular, what is still pending, and to distill a renewed sense of direction to address a complex, at times even intractable, but extremely urgent challenge threatening life at the planetary scale.

A brief introductory note may be helpful to distinguish at least four fundamental tiers of analysis, that correlate to the diverse evolving areas to be considered when aiming to undertake an overall climate change overview, not merely limited to the intergovernmental diplomatic negotiation's purview, seen primarily as a landscape in dynamic transition. Those areas include:

(i) scientific research and assessment in a broad diversity of climate-related fields;
(ii) the diplomatic arena and the international climate policy realm;
(iii) public discourse on global warming;
(iv) the corporate world and the real economy domain, including the carbon-pricing nexus.

Since 2022 at least, an additional fifth level of analysis perforce comprises the issue of climate finance and the examination of the nascent process of reform of the international financial system, its blueprint, and potential transition pathways, even if financial matters have been part of the core issues under negotiations within the climate regime since its inception, and a frequent source of contention in the negotiations along the last three decades. Notwithstanding that, only very recently the possibility of an overarching international financial system reform is being discussed with climate finance overhauling as an attainable goal of the negotiations.

Scientific research comprises a vast body of studies that covers a complex, extensive, and at times even overlapping exploration, in which knowledge of the manifold implications of global warming is being broadened almost ceaselessly, confirming that human-induced climate change is heightening, thus magnifying extreme weather

[1] That sparked a large number of assessments of the results achieved through the sequence of negotiations by a large body of climate researchers on international law, relations, and institutional theory.

events, and aggravating its consequences. In this domain, the Intergovernmental Panel on Climate Change (IPCC) has played a decisive role in systematizing the product of that vast and continuously growing body of research, and given the main impetus to a progressive rise in global awareness. It is now widely accepted as offering the best available scientific foundation for political judgment and decision and provides a scientific narrative that is significant both in international negotiations and in the public sphere, largely, but not only, to support science-based decisions at all levels.

The painstaking construction of a **UN international climate** regime was being conducted in the diplomatic arena, at the multilateral system, primarily by intergovernmental diplomacy under the aegis of United Nations institutions and rules. More recently has widened its interactions and has increasingly extended its linkages with an array of diverse stakeholders, multi-actor coalitions, cities, business leaders, academia, and other non-state actors and institutions. The negotiations have added dimensions to economic strategies, both at the national and corporate levels, extending also to security planning and disaster prevention. In addition, the international climate negotiations—and its intermediate milestones—have played a crucial role in launching and overseeing the largest coordinated scientific cooperation effort (a scientific project as such), in underscoring the equity dimensions of the climate crisis, leading to a call for a just transition, and in supporting the institutional architecture at the national level for developing strategies, action plans, and methodologies for reducing GHG emissions and procedures to adapting to climate change and reinforcing resilience.

The **public discourse** on climate change addresses the challenge to encourage public discussion—grounded in scientific knowledge—to design, develop, and then validate socially and politically viable laws, policies, and measures, and support long-term strategies that should be significant and effective enough to tackle the root causes of the current climate crisis at the relevant scale, and facilitate action towards countries' international commitments, as well as to cope with the increasing risk scenery. In this regard, the narratives and modalities under which the climate change challenge is understood and disseminated are of paramount importance to raise public awareness, enhance societal understanding of risk, and encourage actions by all parties and non-party stakeholders, including civil society, scholars, the private sector, investors, financial institutions, subnational jurisdictions, cities and other subnational authorities, local communities, and indigenous peoples.

Addressing climate change requires a far-reaching transformation of the rules and the entire normative framework under which the global economy typically operated, as well as radically changing the configuration of prevailing patterns of behavior, consumption, and production of contemporary society. The nature, conditions, and architecture of the climate regime necessarily have profound implications on the evolution of the forthcoming **global real economy** in a hyperconnected world. Mutations are to be foreseen in diverse dimensions, including food production and distribution systems, trade and transport systems, connectivity of financial and capital markets, the configuration of global value chains, modalities of competition in international markets, variation of cost structures, the pace of innovation and access to property rights, development and exchange of technology, and communication capacities, as well as in the basic conditions to undertake the transitions towards an almost entirely different energy system, that is the backbone of human activities. However, the aim of this chapter is to examine the challenges ahead, while considering some of the measures and instruments to be utilized to address climate change, such as the prospect for **carbon pricing** to facilitate an efficiently successful path towards net-zero emissions.

In undertaking this analysis, it is also worth looking at the limitations encountered by governmental policies to achieve decarbonization at the level and pace required. Thus, we will refer to the context for international climate policy, to the course corrections currently needed in the climate regime, and called for by a large number of countries and stakeholders, as well as to the longer-term structural reforms to be undertaken to mobilize financial flows through **financial reform** to drive adequate resources towards climate action in mitigation, adaptation, loss and damage, and in associated sustainable implementation investments, in particular to developing countries.

It is worth noting that the transformations to be accomplished in different dimensions to achieve the goals of the Paris Agreement, the Agenda 2030 and in order to protect biodiversity and avoid further transgressing planetary boundaries, are made significantly more challenging in an international geopolitical context in which conflict and fragmentation prevail and are bound to persist or even escalate in the near future.

1.2 Background

A few indicators from different authoritative sources may provide a critical perspective of the magnitude of the challenge that global warming poses to humanity, and may be helpful to signify the extent of the transformations that are needed to tackle this emergency.

Human activities are estimated to have already caused approximately 1.0 °C of global warming above pre-industrial levels.[2] From 1900 to 2020, the world's surface air temperature increased an average of 1.1 °C, due to burning fossil fuels that release carbon dioxide and other GHG into the atmosphere, as well as through other processes, including significant land use change.[3]

If we consider other critical issues, the IPCC estimates that 300 $GtCO_2$ is the remaining carbon budget[4] of global emissions, calculated from the beginning of 2020, and extended until global net-zero CO_2 emissions are reached, if the 1.5 °C

[2] IPCC (2018). Global warming of 1.5 °C. Summary for Policymakers.

[3] UCAR. https://scied.ucar.edu/learning-zone/climate-change-impacts/predictions-future-global-climate

[4] According to the IPCC (2021), "the term 'carbon budget' refers to the maximum amount of cumulative net global anthropogenic CO_2 emissions that would result in limiting global warming to a given level with a given probability, taking into account the effect of other anthropogenic climate

goal is to be achieved (with a likelihood of 83%).[5] In addition, the IPCC estimates that 40 GtCO2 represent the current yearly emissions of CO_2 at the global level. That is, the world has only a few years left, if we are to limit global warming to 1.5 °C, so that we need to ensure a prompt start of that effort. Otherwise humanity is exposed to the differential and growing impacts as "climate models project robust differences in regional climate characteristics between present-day and global warming of 1.5 °C, and between 1.5 and 2 °C" (IPCC 2018). These differences include increases in inter alia: "mean temperature in most land and ocean regions, hot extremes in most inhabited regions, heavy precipitation in several regions, and the probability of drought and precipitation deficits in some regions."[6]

An update issued by the World Meteorological Organization (WMO) in May 2023 indicates that "there is a 66% likelihood that the annual average near-surface global temperature between 2023 and 2027 will be more than 1.5 °C above pre-industrial levels for at least one year. There is also a 98% likelihood that at least one of the next five years, and the five-year period as a whole, will be the warmest on record," thus sounding the alarm that humanity will breach the 1.5 °C level on a temporary basis with increasing frequency.[7]

Looking at these climate issues with a broader perspective, the UN Environment Programme (UNEP) acknowledges that humanity faces when considering climate change, nature, and biodiversity loss, together with pollution and waste, the other two prongs of the daunting triple planetary crisis, as stated in the opening ceremony of the 58th Session of the IPCC by Inger Andersen, the Executive Director of UNEP.

1.3 Negotiations and the Effectiveness of the International Climate Regime

As the construction of the international climate regime evolves, and the need for action increases, and what is more, is further heightened, the analytical perspective inevitably shifts from the focus on those processes aimed primarily at forging a basic consensus on those issues where the urgency and the need to act are considered the most critical, to a new phase. This stage is focused primarily on successfully accomplishing action at the appropriate scale, and achieving the unfolding of socio-technical transitions that are as essential as they seem to be inexorable. That unfolding represents a radical drift from building consensus on selecting priorities, to materializing action at all levels.

---

forcers. This is referred to as the total carbon budget when expressed starting from the pre-industrial period, and as the remaining carbon budget when expressed from a recent specified date".

[5] IPCC AR6 (2021). See Table SPM.2. WGI, the Physical Science Base. Summary for Policy Makers. Page 29.

[6] IPCC (2018). Global warming of 1.5 °C. Summary for Policymakers.

[7] World Meteorological Organization (2023). Accessed in: https://public.wmo.int/en/media/press-release/global-temperatures-set-reach-new-records-next-five-years.

In this regard, it should be recalled that the IPCC has pointed out that the 1.5 °C target would be virtually out of reach, if global greenhouse gas emissions do not peak and reach a maximum in 2025.[8] In fact, the IPCC report, with its plethora of serious concerns and warnings, gives sufficient reasons to understand that course corrections to be adopted now should be significant, and deep, as well as urgent, even if this rationale had been to some extent veiled-through the fractious governmental approval process of the report itself, for political reasons, that are primarily related to national interests, rather than aiming to improve scientific accuracy.

In addition, the momentous Global Stocktake (GST) report,[9] released by the UNFCCC, just on the eve of a sequence of summits held in the run-up to the 28th Climate Change Conference (COP28) into 2024, is unequivocal. The agreement, reached in Paris at COP21, has started to shift the balance, but the actions put in place are not enough. "The world is not on track to meet the long-term goals," this UN report states, in an alarming conclusion.

In fact, the introductory lines in the mitigation section of the GST report are both unequivocal and unsettling: "Global emissions are not in line with modeled global mitigation pathways consistent with the temperature goal of the Paris Agreement," it points out. In addition, the GST highlights that "There is a rapidly narrowing window to raise ambition and implement existing commitments in order to limit warming to 1.5 °C above pre-industrial levels."[10] This wake-up call from the United Nations comes at a time when human-caused global warming has already reached 1.2 °C. Average growth in global fossil CO2 emissions peaked at + 3% per year during the 2000s; in the last decade (between 2012 and 2021) emissions global growth rate slowly declined, attaining a low of + 0.5% per year. However, this slowdown in global growth of CO_2 emissions is still a long way from the massive emission reductions required to achieve the temperature goals enshrined in the Paris Agreement (Friedlingstein et al. 2022, Global Carbon Project).

In addition, global fossil fuel emissions are preliminary projected to raise in 2023 by between 0.5 and 1.5%, while they should have been falling by around 5% in that year, to be consistent with the Paris Agreement mitigation goals.[11]

Despite an array of commitments, pledges, declarations, and national contributions submitted or announced by governments, emissions, indeed, are estimated to have risen again in 2022 and 2023, breaking new temperature and extreme weather records around the world.

Furthermore, the GST report makes clear that "Emissions have peaked in developed and some developing countries, but global emissions have not yet peaked. All Parties need to undertake rapid and deep reductions in GHG emissions in the decades after peaking." The document calls on states to "reduce global GHG emissions by

[8] IPCC AR6 Synthesis Report.

[9] United Nations (2023) Technical dialogue of the first global stocktake. Synthesis report by the co-facilitators on the technical dialogue. Advance version. 8 September 2023. FCCC/SB/2023/9.

[10] See Footnote 9.

[11] Based on projections by the CICERO Center for International Climate Research, data from IEA reports and to be presented within the Global Carbon Project in COP28.

43% by 2030 and further by 60% by 2035, compared with 2019 levels"; those levels of reductions being the only way to stay within the boundaries of the Paris Agreement and making it plausible to achieve carbon neutrality by no later than 2050.

There is, therefore, a daunting set of gaps that persist, even increase, both between the level of ambition of the commitments already made by the Parties to the UNFCCC, and what needs to be effectively achieved to respond to global climate change—in line with the Paris Agreement—to preserve the sustainability of life on the planet. Similarly, there is a huge gap between those commitments and their effective implementation, which is at the very least drawn-out, if not directly stalled.

In addition, those gaps should include those referred to the issues of financing[12] and of the amount of funds needed and available, which in turn present a stark imbalance between the different dimensions of institutional response being undertaken. That response is in itself intensely asymmetric in its distribution between mitigation, adaptation, and effectively addressing loss and damage.

The inability or reluctance to act has significantly aggravated the global situation, to the point that further delays in mitigation may seem almost as intolerable, to the extreme that they may even jeopardize governance in some highly vulnerable countries. Thus, the emphasis on implementation had emerged, prior to the COP28 summit, as a central point to be considered where concrete and meaningful results were expected.

1.4 On the Nature of International Climate Governance

It has been posited that a legally binding, multilateral agreement is a necessary condition for achieving the highest levels of GHG emission reductions, consistent with limiting warming and pursuing net-zero emissions, in line with the mitigation goals of the Paris Agreement. Typically, it is understood that clear legally binding commitments, within a multilaterally agreed process, with strong legal and institutional characteristics, are needed to give countries the confidence that their national interests, in particular the economic ones, are being fairly and equally treated (Hare et al. 2010; Flachsland 2015), and that efforts to ensure a just transition are part of the tenets of such an agreement.

Thus, the adoption of the Paris Agreement, in December 2015, was portrayed by the then United Nations (UN) Secretary-General Ban Ki-moon as "a resounding success for multilateralism" (UNFCCC 2015). However, although the UNFCCC governance regime can be viewed, at least initially, as a form of "monocentric" governance, in the more than three decades of intergovernmental efforts to address climate change, it has become increasingly clear that it operates rather as part of an evolving polycentric governance realm.

[12] As well as the provision of other means of implementation.

This evolution was interpreted by researchers under a spectrum of terms that include "fragmentation" of international institutions (Biermann et al. 2009; Zelli 2011), having the entity of a "regime complex for climate change" (Victor and Keohane 2010), even as "experimentalist governance" (Sabel and Victor 2017), or as a "reconfiguration of authority in global climate governance" (Hickmann 2017).

By the late 2000s, Elinor Ostrom was elaborating an innovative positive narrative: not all aspects of governance would have to be painstakingly designed by international negotiators in endless sessions, at times reveling in regulatory micromanagement. New forms were, she sustained, emerging almost spontaneously from the bottom-up, producing a more decentralized and multilevel pattern of governing, which she prefers to call "polycentric."

When exploring the challenges to be faced in achieving sustainability in socio-ecological systems and dealing with a classic collective action problem, such as GHG emissions and, from an overall perspective, global warming, Ostrom foresees a pathway for the future of the commons, that goes beyond the adverse dichotomy of market failure and complicated government regulation. As an alternative, she proposes a polycentric approach at various levels, with the oversight of local, regional, and national stakeholders. The efforts to reduce global GHG emissions, Ostrom underscores, are best addressed at multiple scales. Given the slowness and widening "conflict involved in achieving a global solution to climate change, recognizing the potential for building a more effective way of reducing GHG emissions," can be considered to be an important step forward (Ostrom 2009).

According to van Asselt and Zelli (2018), the international climate regime has been able to exert at least some influence on the emergence of national, private, subnational, and transnational climate governance. But how much influence it was able to impose during that process or is now wielding—especially compared to other possible driving factors of action, such as the quest for competitiveness, moral concerns, gaining co-benefits, protecting investment or industrial sectors, the influence of transnational corporations, or avoiding leakage, and through precisely what causal mechanisms still remains unclear or disproved.

It is helpful to look at the models of international law, as applied in climate change negotiations, to elucidate which one of three different models may provide the best response to the dilemmas posed by global climate change. Bodansky (2023) identifies three models of international law and examines their capability to address a complex collective action problem. These models are:

(a) a prescriptive model;
(b) a contractual model; and
(c) a catalytic model.

Again, according to Bodansky, the prescriptive model raises issues of legitimacy and effectiveness given the central role of state consent in international law, while "strategies that involve using international law to compel states to change their climate change policies seem unlikely to prove successful" even in the best of circumstances. Experiences gained in the process of negotiations in the recent past indicate that this assessment may be at least partly right. Due to frustration and disappointment with the slow pace of the negotiations, the actual or anticipated risks of climate change impacts, and the increasing severity of extreme events the world over, the process gave rise to a number of proposals by researchers to address climate change through international adjudication, in fact supporting the notion of international climate litigation and its proliferation (Bodansky 2023).

Then, under the contractual model the assumption is that states may enter into treaties because they assume treaties leave them better off, by a reciprocal exchange of commitments. However, also according to Bodansky, the contractual model conceptualizes the climate change dilemma in the wrong way, as it involves outstanding issues of domestic policy and has huge implications for a country's economic sovereignty; further, it is complex and the outcomes are frequently uncertain, and it results in domestic winners and losers. Hence, progress in addressing the climate change issue and reaching significant GHG emission reductions is driven much more by domestic politics issues than by international politics (Bodansky 2023).

The virtual stasis of the Kyoto Protocol lead to embracing a different international law model, the catalytic model, after four years of negotiations, from 2011 to 2015,[13] and adopting the Paris Agreement that aimed to better reflect contextual changes, new perspectives, and a reconfiguration of authority, while still reaffirming the centrality of the UNFCCC process, albeit including new stakeholders that are relevant actors in global affairs.

When referring to the catalytic/facilitative model, the third of the alternatives of international law models identified before, it is worth noting that the main purpose of a catalytic regime is to shift actors' preferences and strategies towards cooperative outcomes over time, thus facilitating action, primarily focused at promoting deep transitions and changes at the national level.

Examining the dilemmas of cooperation, Hale (2020) seeks to explain how and through which conditions, international institutions facilitate cooperation between states. In that endeavor he distinguishes mainly two approaches: the regulatory model that comprises both the prescriptive and contractual law models, and the catalytic and facilitative model.

The lack of incentives to implement their proposed policies made states and other actors attempt to solve the problem, through the creation of catalytic institutions, in a process of innovation and adaptation of the regime to an increasingly adverse context, going beyond the polycentric structures in place, or more precisely, shifting from the logic that underlined the Protocol.

If we examine the evolution of the formal configuration of the overall process of negotiations towards the completion of a UN international climate regime, we can single out at least two different shifts that reflect the substantial evolution of the international climate architecture, that adapted in order to enhance climate action in a world ever more reluctant to implement mitigation actions. Those shifts implied moving from:

(a) the universal and monocentric architecture initially adopted, to an essentially polycentric approach, more nuanced and increasingly pluralistic;

[13] After the failure of the Copenhagen Accord and the attempt in Cancun to restore trust.

(b) from the endorsement of a prescriptive international law model, to embracing a catalytic/facilitative model, in an effort to reflect the changing conditions and constraints in the context in which negotiations were unfolding, while preserving or even reinforcing the logic of cooperation.

These contemporaneous shifts took place in the course of the last three decades, while the UN climate change regime evolved, starting with the inception of the UNFCCC in 1992, through the efforts made under the Kyoto Protocol, since its approval in 1997, and the provisos included in the Paris Agreement, starting with its entry into force in 2016.

In addition, when keeping track of the changes in the institutional and operational settings in terms of the effectiveness, efficiency, and equity of the architecture in place for meeting the Convention's objectives, through the prism of a substantive perspective, it is possible to distinguish three different passages:

(i) the move away from targeting developed country emissions, without imposing quantitative requirements for emissions reductions (under the Convention), only a time-related cap; to establishing quantified emission reductions for industrial countries in a legally binding manner (under the Kyoto Protocol), and then aiming at universal participation by requiring voluntary contributions and hence action from all countries, under the Paris Agreement;
(ii) the shift from the top-down Kyoto architecture to the hybrid Paris paradigm, with bottom-up approaches; and
(iii) the expansion from a solely mitigation focus under Kyoto, to a multiple complex goal setting comprising: mitigation, adaptation, and finance under the Paris Agreement.

Much of the discussions of the future of the international climate change regime at the policy design level, as well as in the field of research on international relations and institutions, is concerned with changing patterns of authority, and its implications in very different domains. The viewpoints of policymakers and scholars naturally vary, and there are in both fields' different interpretations and reasons for concern. If we classify those reasons for concern, there are basically three distinct classes, as follows:

(i) failure to deliver, related to efficiency and effectiveness.
(ii) erosion of sovereignty, related to equity, transparency, and fairness.
(iii) fragmentation.

In each case, there is a multiplicity of nuanced concerns that are associated with a set of dimensions that may be related, or even somewhat overlap in a few circumstances.

Failure to deliver	There are different issues, of a different nature and relevance, attributed to the climate regime architecture, currently the Paris Agreement, or its limitations: • Mitigation gap at the global level: "global emissions are not in line with modelled global mitigation pathways consistent with the temperature goal of the Paris Agreement" (GST, KF4); hence the mitigation goal may not be achieved. There is both a quantitative dimension (not enough reductions) and a temporal dimension (reductions are not achieved within the time limits established as global goals) • Emission gaps at the country level: "Based on current NDCs, the gap to emissions consistent with limiting warming to 1.5 °C in 2030 is estimated to be 20.3–23.9 Gt CO_2 eq. 2" (GST KF4.10) • Implementation gaps: "Urgent action and support are needed to ramp up implementation of domestic mitigation measures" (GST KF5.14) • Insufficient ambition: "More ambitious mitigation targets in NDCs are needed to reduce emissions more rapidly" (GST KF5.15) • Insufficient adaptation: "Most observed adaptation efforts are fragmented, incremental, sector-specific and unequally distributed across regions" (GST. KF 10) • Mitigation bias: "The share of adaptation in mobilized climate finance increased—from 20 percent in 2017–2018 to 28% in 2019–2020" but it is still less than a third of total climate finance available (GST KF 14.49) • Finance gaps: "It is essential to unlock and redeploy trillions of dollars to meet global investment needs, including by rapidly shifting finance flows globally to support a pathway towards low GHG emissions and climate-resilient development" (GST KF 15.50)
Erosion of authority[14]	• Outsourcing of global governance • Decline in accountability and trust in the international community • Shift in implementation from OECD governments to TNCs based in OECD countries • Serving particular interests These concerns emerge particularly in the context of North–South dynamics and contentious perspectives
Fragmentation	• Coordination gaps • Regulatory uncertainty

Source Own elaboration
Note: *GST* Global stocktake, *KF* key finding

1.5 The Road Ahead

1.5.1 The Near-Term Perspective

In more than one relevant dimension, it can be said that the climate negotiations presently under the Paris Agreement are now effectively at the halfway point. On the one hand, seven years have already elapsed since the momentous signing of a

[14] Gleckman (2023).

landmark Paris Agreement, and there are almost another seven years to go until 2030, which is a critical milestone to consolidate substantive momentum on both mitigation and adaptation action, building resilience as well as in sustainable action towards the 2030 Agenda.

In addition, the Sustainable Development Goals (SDG) Summit and the Climate Ambition Summit held in September 2023, mark the halfway point to the 2030 Agenda and should aim to provide renewed impetus for accelerating the pace of change in the field of sustainable development, addressing climate change, and protect biodiversity.

It is therefore necessary in the first place to focus the dialogues that are to take place during this incoming period and continue in the near future to seek a common ground, especially around three particularly important persisting asymmetries, so that the interventions that will be made in the next incoming negotiation sessions will give meaning to and effectively consolidate the blueprint for the robust course corrections that the next sessions of the Conference of the Parties should intend to elaborate and decide upon.

1.5.1.1 Decarbonization

One of the key issues to be agreed upon in the short run is how to effectively accelerate the decarbonization process, since a question raised by stakeholders is whether governments will fulfill the commitments they have made, through indecisiveness or procrastination.

In the first place, it is necessary to either begin, broaden, or (in the case of some countries that have strengthened their climate policies the most) simply deepen the decarbonization processes already underway. In this regard, the UN is putting the issue of phasing out fossil fuels back at the core of the COP agenda.

It is vital to contemplate, in a general way, achieving a downward inflection point in the current emissions curve, through an ensemble of emission reduction agreements that push global emissions downwards. Simultaneously, global targets aimed at accelerating and strengthening the transition away from fossil fuels towards expanding renewables could be simultaneously agreed, as an essential component of a truly 1.5 °C-aligned just energy transition, while the expansion of coal, oil, and gas is incompatible with a that 1.5 °C energy pathway (IEA). More aggressive methane reductions should be part of the efforts to be made.

These mitigation priorities should comprise substantially more ambitious emission reduction targets as well as significantly more robust systems, so that relevant stakeholders (regions, countries, states or provinces, cities, public and private companies, corporations, small- and medium-sized enterprises, partnerships, etc.) can appropriately and credibly account for both the commitments they make going forward, as well as the extent to which they are being able to meet the mitigation pledges they have already made.

In addition, the actual implementation of mitigation policies, particularly in developing countries, faces a major obstacle, due to the high levels of indebtedness of a large number of countries and the high cost of capital in some of these countries, which acts as a significant and persistent barrier and discourages the implementation of plans, policies, and measures, and realizing investments aimed at mitigating climate change on the appropriate scale together with choosing the optimal technology options available.

1.5.1.2 Resilience

It is noticeable that while some mostly incremental efforts to escalate mitigation are being made, the climate crisis has deepened more acutely, indeed at an alarming rate. In this last period, there has been almost daily evidence of the increasingly severe and at times irreversible consequences that are emerging while the planet is forced to cross the threshold of 1.5 °C of warming, compared to pre-industrial times, including catastrophic floods in Pakistan and China, record heat waves in the USA and Europe, extremely severe droughts in Africa, record melting of the Poles in 2022, fires in Canada and USA, and severe and widespread droughts in Latin America, among other extreme events, that persisted or in some cases were even aggravated in 2023.

Millions of people are already exposed to acute climate change-driven insecurity of access to and availability of food and water, while progress in adapting to the impacts of global warming is uneven, fragmentary, and certainly insufficient to prevent human suffering and loss of life. The condition of vulnerability to impacts, from which almost no stakeholder escapes, is aggravated by the uncertainty about risks, insufficient data or the incomplete nature of the information available, in particular in developing countries, as well as the rapid evolution of circumstances, which includes climatic issues, but also fluctuations in impacts and the progressive and relentless deterioration of biodiversity on a planetary scale.

Progress in adaptation is forcefully constrained by the lack of financial resources and access to appropriate technologies, but also to the scarcity of technical and organizational capacities, which makes it even more difficult the process of adaptation in a context of conflicts partly unleashed by the adverse effects themselves, with their aftermath of social and even political disruptions, in the case of extreme events with catastrophic characteristics, such as some of those mentioned above.

1.5.1.3 Financing

The matter of financing emerges as an essential issue to make climate action viable on a global scale; it is critical, has been long discussed, and its resolution almost always postponed for very different reasons related to politics, fractious exercise of power, conflicting interests, or the prevalence of stringent fiscal policy approaches even in those countries most affected.

Finance is a pillar of transformation, making it possible, whether in the direction of mitigation or to reduce vulnerability to impacts and strengthen the adaptive capacity of societies, natural systems or indeed in productive assets.

However, after three decades of debate, progress in climate finance under the UNFCCC has certainly been limited in substance, despite the proliferation of institutions (through the multiplication of funds with different designations, institutional arrangements, and anchors) created to manage financial resources that are not materialized in the necessary magnitude, timeliness, predictability of provision, and appropriate conditions of access.

The only way, it must be stated, to ensure that the reduction of GHG emissions—at the speed and in the volume called for by the IPCC—becomes a reality throughout the planet, and particularly in developing countries, is to ensure the provision of climate funds on a radically different scale, congruent with the financing requirements in those countries and their investment needs, especially in addressing infrastructure demands. The latter, according to UNCTAD, amount to USD 4 trillion in developing countries, if the Sustainable Development Goals are to be achieved by 2030, in line with the 2030 Agenda aims.[15]

In short, what must be considered, agreed upon, and integrated with other ongoing efforts with similar purposes is a comprehensive reform of the international financial system to bring it into line with the Paris Agreement, with the existence of a carbon budget that rapidly tends to zero, and in the limit, with the adoption of new business models that reflect the current risky planetary conditions and tend to avoid their worsening, as well as their more detrimental effects.

On the other hand, developing countries (and the emerging markets that still negotiate within the G77 + China Group) account for an increasingly larger share of global emissions, currently accounting for 64.9% of total GHG emissions, if China's emissions (26.1%) are included, while most of the capital investment needs are located in the developing countries, and are necessary to transform their markets and their existing infrastructures.[16]

The future decisions of developing countries in terms of infrastructure (expansion or renovation), together with productive transformation, will therefore have a significant influence on the expected climate risk on a global scale. The process of capacity building to mitigate risk would not allow these structural risks to be addressed immediately, nor are there always in place the financial structures, mechanisms, or financial engineering instruments available in the private sector that would allow risk diversification on the appropriate scale and with the necessary urgency.

Moreover, according to a number of large-scale global fund managers, the current involvement of private investors in climate finance in emerging markets is relatively limited. Although private capital is theoretically abundant, given the relatively poor investment conditions in a number of developing countries, even in the larger emerging markets, the reality is that private funds will not flow in sufficient magnitude

[15] UNCTAD.ORG. World Investment Forum (2023). Accessed in: https://worldinvestmentforum.unctad.org/session/high-level-roundtable-investment-sustainable-infrastructure.

[16] World Resources Institute. GHG Emissions 2018, excluding land-use change and forestry.

to developing countries (with perhaps a few exceptions in a number of very specific markets), unless there are in place new risk-sharing mechanisms, appropriate regulatory frameworks, and the additional contribution of greater public sector exposure to absorb eventual losses.

1.5.2 The Long-Term Perspective

1.5.2.1 The Institutional Architecture

As a consequence of the concerns related to the inadequate functioning of the regime and those associated with widely held assumption that the emergence of transnational governance arrangements generates a general shift of authority from the intergovernmental level towards sub-state and non-state actors. This shift could be understood as an erosion of sovereignty as a direct consequence of the inadequacy of the regime to perform its functions and achieve its goals.

Conversely, we observe that the ongoing process is one of reconfiguration of authority that apparently reinforces the importance of state-based forms of governance, while seeking effectiveness, broader participation, and expand capacity to address an increasingly complex process with implications, linkages, and interrelations with a multiplicity of domains.

We envisage that the UN international climate regime will remain significant as a core domain safeguarding the centrality of the international climate regime clustered around the United Nations Framework Convention on Climate Change for delivering "key governance functions that are essential to the functioning of the overall system,"[17] including the following:

(i) ensuring adherence to sound ethical and political guiding principles of the international climate regime;
(ii) establishing overarching norms, rules, and decision-making procedures;
(iii) coordination of research, in liaison with the IPCC, as well as overseeing that research priorities, triggered by longer-term politically strategic objectives, are reflected appropriately in further scientific efforts;
(iv) ensuring that transparency prevails in all process-related issues and, in particular, in the system of reporting and review;
(v) sorting through the many methodological and technical issues and verifying the coherence of its design and application;
(vi) supporting national efforts associated with climate change action, in adaptation and mitigation, and loss and damage, as well as in complementary outreach activities;
(vii) assisting in the process to reform the international financial architecture in what relates to climate finance and sustainability-related matters;

[17] Betsill et al. (2015), Page 3.

(viii) spur increasing ambition in climate action and support;
(ix) supporting the consideration of other relevant matters, such as trade-linked implications of new policies and measures, and other emerging issues;
(x) ensuring that equity, inclusion, and just transitions are considered throughout the entire architecture of the governance regime.

Even if these roles seem inordinately comprehensive, after changes undergone, they primarily relate to ensuring compliance with the guiding principles and safeguarding transparency, consistency across the present constellation of dominions, and connectivity with other international regimes in related matters, such as international trade issues. In addition, norms, rules, and procedures established at the intergovernmental level ought to have a significant impact on the initiatives launched by subnational and non-state actors (see Figure).

1.5.2.2 Macroeconomic and Financial Policy Tools for Mitigation Towards Net-Zero and Sustainability

It has repeatedly been alleged that the window of opportunity to stop global warming is closing rapidly, in spite of the many countries that are announcing and starting to pursue more ambitious policies and measures to reduce GHG emissions. Simultaneously, adaptation needs are huge in many of the most vulnerable countries, where fiscal space is limited, tax capacity enforcement is weak, and access to international finance is almost nil. These conditions lead to the consideration of a nuanced mix of policies to be able to achieve the goals of the Paris Agreement.

It is worth noting that action is springing up in major nations and in different regions, aiming to achieve net-zero GHG emissions within a few decades, in line with the Paris Agreement mitigation goals. Shifts towards renewable energy are taking place in diverse sectors, but not at the pace needed. Carbon pricing through taxes, tariffs, tradeable permits, as well as the slow removal of fossil fuel subsidies and the introduction of incentives for renewable energy and carbon sequestration (i.e., CCUS techniques) are increasingly implemented at larger scales, albeit still insufficient to make emissions peak in the immediate future.

There are substantial material, societal, and emission gains to be made from transforming consumption and production patterns, renovating infrastructure to reinforce resilience, and making progressive shifts towards a circular economy. Notwithstanding that, there is a large consensus that the pace of these promising changes must urgently increase in order to reach the global climate mitigation goal. A wide range of macroeconomic policy tools can tackle climate change and can be part of the package of measures to ensure mitigation. An extensive spectrum of fiscal policy tools is available. Fiscal instruments comprise the following categories:

- price policies (taxation, subsidies);
- spending and investment; and
- public guarantees.

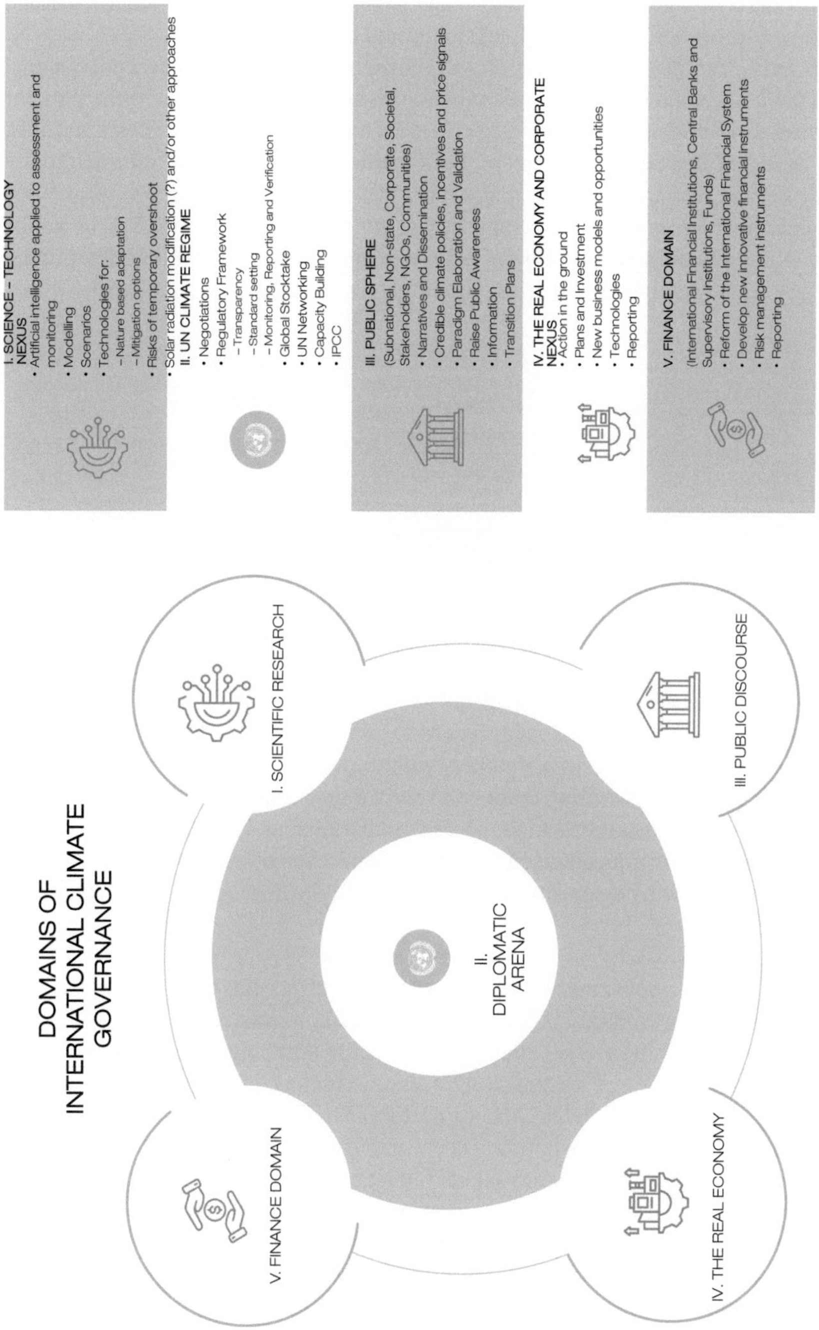

Source Own elaboration

Price policies include inter alia carbon taxation and cap-and-trade schemes, subsidies for introducing green technologies and fostering mitigation action, low-carbon investment subsidies, interest rate subsidies, and tax breaks.

Spending and investment tools include outright public investment, concessional loans from development banks, and public investment funds. Several of the policies and measures adopted by a number of countries rely strongly on spending measures, such as increasing public investment to facilitate transitions through infrastructure expansion and renovation. Yet, in many cases, these policies entail heavy fiscal costs.

Public guarantees can help secure higher private sector participation in public financing as it could play a critical role in reducing risk perceptions for project developers, financial intermediaries, and investors and minimize fiscal cost.

Fiscal	Financial	Monetary
• Carbon pricing • Carbon tax • Cap-and-trade • Tax relief • Tax reductions • Tax exemptions • Public spending • Public investment • Social spending • Climate funds • Other • Debt swap	• Cost of capital reduction • Concessional loans • Risk mitigation/transfer • Loan commitments • Guarantees • Insurance • Capacity building • Grants • Risk disclosure • Taxonomy and policies	• Credit allocation • Green banks • Indirect credit easing • Green targeted refinancing operations • Integrate climate risks • Risk assessment • Differentiated collateral valuation

Source Based on Krogstrup and Oman (2019)

Climate action presents policymakers with difficult tradeoffs. Relying mostly on spending measures and scaling them up to deliver on climate ambitions will become increasingly costly, according to the IMF, possibly raising debt by 45–50% of gross domestic product by midcentury. High debt, a landscape of rising interest rates, and weaker global growth prospects will further make public finances harder to balance.[18]

Carbon Pricing

"Carbon markets—both emission trading systems and baseline and credit systems—are an increasingly common policy instrument being introduced to address climate change mitigation" (Betz et al. 2022). Further, carbon pricing, through a carbon tax or an emission allowance system, might be an efficient instrument to incentivise energy transitions and growth. According to J. Stiglitz 2020 although there is a large body of economists who consider a carbon-pricing policy to be necessary to accomplish ambitious CO_2 emissions reductions in large economies, most such economists would also recognize a carbon-pricing policy will not be sufficient.

Public investment—government financing and more direct action through public works—are necessary in the context of a broader societal transformation, as well as expected changes in patterns of behavior.

[18] International Monetary Fund (2023).

Hence, on their own, markets cannot deliver sufficient mitigation. "Market failures, unaddressed and exacerbated by government failures, prevent an appropriate market only response to the challenge of mitigating climate change" (Krogstrup and Oman 2019).

It is worth noting that carbon markets have given rise to regulatory challenges since their inception, yet progressive policy responses to these challenges have grown in sophistication over time. Regulatory challenges in these markets will continue to evolve, under the frameworks, envisaged in Article 6 of the Paris Agreement, as the value of traded units would gradually increase, and if markets are able to become more interconnected, whether through bilateral linkage or multilateral cooperation. New digital tools, such as software-based analysis of market data and remote sensing for compliance control and monitoring, will allow to enhance the accuracy of MRV procedures (Betz et al. 2022). At any rate, if carbon markets are to play a role in advancing the global effort to decarbonize our economies robust and sensible regulation will only increase—not lose—in importance going forward.

Box: Carbon Pricing in Latin America and the Caribbean Carbon prices used in Latin America and the Caribbean are not quite enough yet to reduce emissions, at scale, stimulate low-carbon investment, and fund emission captures and stimulate technological change, increase revenues, and mobilize private investment. One reason carbon prices are difficult to implement is that they might imply high additional costs on poor and vulnerable households.

While in theory carbon prices can stimulate emission reductions, in practice even if utilized worldwide, have not had a significant impact yet. According to the IADB, emissions reductions from existing carbon prices are limited to between 0 and 2% per year. In addition, the evidence suggests that carbon prices incentivize incremental changes but do not result into transformational investments funding the technology and infrastructure required to reach net-zero emissions. There are at least two causes for this limited impact: firstly, carbon prices frequently are not high enough and do not comprise all sectors. In the second place, governments should act on regulations and reinforce infrastructure deployment to enable the required transitions. In addition, carbon pricing alone cannot fund the key transitions, which will require an estimated spending of around 7–19% of GDP every year on climate-related goals while at the macroeconomic level, proceeds from carbon prices are still very small. Due to the combination of low taxation and high fuel subsidies, the region collects the second lowest tax revenue from fossil fuels in the world: just 0.06% of GDP, compared to the world's average of 0.7% (IADB).

"Despite the challenges associated with carbon prices, they can still be helpful if well-designed, backed by strong institutional capacities, and used with realistic objectives within a mix of environmental policies."[19]

Voluntary carbon credit markets may be another possible funding source. But designing credible credit markets is not an easy task; it requires accurate monitoring, reporting, and verification systems. Moreover, the size of the voluntary carbon credit markets is still insufficient. In 2021, voluntary markets generated almost US$2 billion globally, whereas the region needs to invest $280 billion every year just to provide basic access to decarbonized and climate-resilient infrastructure services. In addition, in 2023 serious doubts on the integrity of a mega-project in the 2 billion carbon offsets market have put the market in a virtual gridlock.

Industrial Policies

Industry is closely connected to climate change: energy use in industry accounts for 24.2% of global greenhouse gas (GHG) emissions, while the share of GHG emissions from direct industrial processes is 5.2%. Policy action for greening industrial development is increasingly being recognized as a key element of the transition to a low-carbon future (Anzolin and Lebdioui 2023).

A green industrial policy faces several challenges, including reinforcing green manufacturing of the equipment required for net-zero transformation (i.e., wind turbines, solar modules, hydrogen electrolyzers heat pumps, batteries); supplying and generation of green energy (i.e., electricity from wind/solar power, clean hydrogen, or derivatives); and green production carbon–neutral production of industrial products (green steel, metals, cement, glass, chemicals production). In addition, further challenges are related to the instruments used to propel the transformation approach as an increasing number of countries base their transformation policies on subsidies rather than carbon pricing. For example, the US Inflation Reduction Act (IRA) poses a challenge to the EU by providing substantial subsidies for the development of green value chains in the form of transparent and easy-to-understand tax credits.

At the policy level, an increasing number of governments are promoting new industrial policies that might enhance the role of state intervention, and comparatively diminish that of carbon pricing, in the context of a fragmented world economy and geopolitical turbulence, where global value chains might become relatively less significant, due to ideological divides and competition for technological preeminence in a number of key technologies and value chains.

Transition Plans

Private sector transition plans can play a vital role in accelerating a consistent transition to a climate-resilient world, help to minimize risks, as well as alleviate the potential high up-front costs of socio-technical transitions, together with reducing

[19] Talbot Wright et al. (2023).

disruptions in the functioning of the real economy, precisely as a result of those transitions.

Transparent climate transition plans are an important mechanism for organizations to demonstrate to investors, consumers, and other stakeholders that they have developed a strategy that will keep their business on the pathway to 1.5 °C and are thus able to remain profitable in the context of a net-zero economy.

Meanwhile, it is essential to ensure that private sector transition plans be aligned with, and contribute to, their respective country's NDC, and support national or regional transformation pathways, that could also help resolve the tensions between the needs of different jurisdictions.

Final Considerations on the International Context

There are a number of conclusions that can be made after examining the state and perspectives of the UN international climate regime, in a succinct manner. Firstly, climate change is an extremely complex challenge and a number of scholars have referred to the wicked nature of the problem it poses. The UN international climate regime has achieved some progress in the thirty years of multilateral negotiations that unfolded, but much remains to be done and time is of the essence. The UN climate regime evolved, trying to adapt to international circumstances, while urgency for action heightened as impacts increased and understanding of risks deepened continuously.

The combination of concerns over failure to address mitigation at the necessary scale and within the time boundaries expected created the need to find more effective arrangements by shifting the architecture of the regime and the nature of the commitments required from states. Simultaneously, as climate change scaled in the international political agenda and citizens, the world over increased their awareness of the implications of global warming, and other environmental disruptions resulting from transgressing key planetary boundaries, in a way as a positive effect of work done under the UN climate regime, the demands for effective action expanded beyond governments in a network of key stakeholders, civil society, academy, and dedicated NGOs calling for more participation, effectiveness, transparency, and enhanced outcomes.

The Paris Agreement was then seen as an important step forward on the path to a low-carbon economy and a resilient society, in a new negotiating phase. However, Paris paradigm represented what was seen as a shift of authority—in the limit, a fragmentation of authority—that elicit concerns on the preeminence of transnational climate initiatives leading eventually to transnational climate governance (subnational, nonprofit or corporate non-state), that might erode governmental authority and national sovereignty and emerge as alternative governance arrangements to those created by national governments along the negotiation process. This is seen, critically, as a relocation of authority from state-based forms to new actors and institutions as well as an erosion of authority that is risky for developing countries.

Conversely, the ongoing reconfiguration of authority tends to consolidate the importance of state-based forms of governance, given the challenging geopolitical context in which the fight against climate change is being had. Hence, principles, norms, rules, and decision-making procedures, methodologies, and transparency requirements established under multilateral settings are essential for the functioning of the overall climate regime. No other arrangement would then be viable in the increasingly fragmented international geopolitical regime, increasingly divided, and a global economy that seems to be progressively more segmented, limiting the possibility, for example of agreeing in a global carbon price and trade relations, among other key issues.

If this interpretation is correct, then the global response to climate change is essential to establish once again the centrality of the UN climate regime to provide the norms and rules, the overall framework, agreed upon by nation-states in international climate negotiations, under which different stakeholders would operate and can hence be regarded as complementary to the UNFCCC process, combining consistency and effectiveness, through the expansion of the network of stakeholders at all levels, and flexibility to ensure climate action (adaptation and mitigation) at the scale and pace needed, through a multiplicity of actors in line with the notion of universal participation. Against these advantages, there is the risk of erosion of sovereignty resulting from the asymmetry of power between diverse stakeholders that could hamper the process of attaining just transitions and equitable outcomes. Hence, there is a need to safeguard transparency and put in place rigorous procedures to measure results obtained and the overall integrity of the processes established to address climate change.

References

Anzolin G, Lebdioui A (2023) Policy action for greening industrial development: coordinating low carbon consumption, production & innovation. Policy brief series: insights on industrial development. UNIDO. Issue NO. 5–July 2023

Betsill M, Dubash N, Paterson M, van asselt H, Vihma A, Winkler H (2015) Building productive links between the UNFCCC and the broader global climate governance landscape. Glob Environ Polit 15(2):1–10

Betz R, Michaelowa A, Castro P, Kotsch R, Mehling M, Michaelowa K, Baranzini A (2022) The carbon market challenge. Cambridge University Press, Cambridge. https://doi.org/10.1017/9781009216500

Biermann F, Pattberg P, van Asselt H, Zelli F (2009) The Fragmentation of global governance architectures: a framework for analysis. Glob Environ Polit 9:14–40

Biermann F, Hickmann T, Sénit C (eds) (2022) The political impact of the sustainable development goals: transforming governance through global goals? Cambridge University Press, Cambridge. https://doi.org/10.1017/9781009082945

Bodansky, D.(2023). The UN Climate Change Regime Thirty Years on: A Retrospective and Assessment. Environmental Policy and Law 53 (2023) 19–33.

Cole D (2015) Advantages of a polycentric approach to climate change policy. Nat Clim Chang 5(2):114–118

Depledge J, Yamin F (2009) The global climate-change regime: a defence. In: Helm D, Hepburn C (eds) The economics and politics of climate change. Oxford University Press, Oxford, pp 433–453

Fay M, Hallegatte S, Vogt-Schilb A (2013) Green industrial policies: when and how (English). Policy research working paper; no. WPS 6677. World Bank Group, Washington, D.C.
Flachsland C (Prof. Dr.) et al (2015) The architecture of the global climate regime. https://doi.org/10.3763/cpol.2010.0161
Friedlingstein P, O'Sullivan M, Jones MW, Andrew RM, Gregor L, Hauck J, Le Quéré C, Luijkx IT, Olsen A, Peters GP, Peters W, Pongratz J, Schwingshackl C, Sitch S, Canadell JG, Ciais P, Jackson RB, Alin SR, Alkama R, Arneth A, Arora VK, Bates NR, Becker M, Bellouin N, Bittig HC, Bopp L, Chevallier F, Chini LP, Cronin M, Evans W, Falk S, Feely RA, Gasser T, Gehlen M, Gkritzalis T, Gloege L, Grassi G, Gruber N, Gürses Ö, Harris I, Hefner M, Houghton RA, Hurtt GC, Iida Y, Ilyina T, Jain AK, Jersild A, Kadono K, Kato E, Kennedy D, Klein Goldewijk K, Knauer J, Korsbakken JI, Landschützer P, Lefèvre N, Lindsay K, Liu J, Liu Z, Marland G, Mayot N, McGrath MJ, Metzl N, Monacci NM, Munro DR, Nakaoka S-I, Niwa Y, O'Brien K, Ono T, Palmer PI, Pan N, Pierrot D, Pocock K, Poulter B, Resplandy L, Robertson E, Rödenbeck C, Rodriguez C, Rosan TM, Schwinger J, Séférian R, Shutler JD, Skjelvan I, Steinhoff T, Sun Q, Sutton AJ, Sweeney C, Takao S, Tanhua T, Tans PP, Tian X, Tian H, Tilbrook B, Tsujino H, Tubiello F, van der Werf GR, Walker AP, Wanninkhof R, Whitehead C, Willstrand Wranne A, Wright R, Yuan W, Yue C, Yue X, Zaehle S, Zeng J, Zheng B. (2022) Global carbon budget 2022. Earth Syst Sci Data 14:4811–4900
Gilligan JM, Vandenbergh MP (2020) A framework for assessing the impact of private climate governance. Energy Res Soc Sci
Gleckman H (2023) Multistakeholderism: is it good for developing countries? South Center–Transnational Institute
Guofeng Wang & Changpeng Huan (2023) Negotiating climate change in public discourse: insights from critical discourse studies. Crit Discourse Stud. https://doi.org/10.1080/17405904.2023.2198725
Hale TN (2020) Catalytic cooperation. Glob Environ Polit 20(4):73–98
Hare W. (2010), Stockwell, C., Flachsland, C., & Oberthur, S. (2010). The architecture of the global climate regime: a top-down perspective. Climate Policy, 10(6), 600–614. https://doi.org/10.3763/cpol.2010.0161. Glob Environ Polit 16:12–22
Hickmann, T. (2017). The Reconfiguration of Authority in Global Climate Governance. International Studies Review 19(3):430-451
International Monetary Fund (IMF) (2019) Macroeconomic and financial policies for climate change mitigation: a review of the literature. IMF Working Paper. Washington, DC: IMF, September 2019
International Monetary Fund (IMF) (2023) Fiscal monitor: climate crossroads: fiscal policies in a warming world. IMF, October, Washington, DC
IPCC (2018) Summary for policymakers. In: Masson-Delmotte V, Zhai P, Pörtner H-O, Roberts D, Skea J, Shukla PR, Pirani A, Moufouma-Okia W, Péan C, Pidcock R, Connors S, Matthews JBR, Chen Y, Zhou X, Gomis MI, Lonnoy E, Maycock T, Tignor M, Waterfield T (eds) Global Warming of 1.5°C. An IPCC special report on the impacts of global warming of 1.5 °C above pre-industrial levels and related global greenhouse gas emission pathways, in the context of strengthening the global response to the threat of climate change, sustainable development, and efforts to eradicate poverty. World Meteorological Organization, Geneva, Switzerland, 32 p
Jernnäs M, Lövbrand E (2022) Accelerating climate action: the politics of Nonstate actor engagement in the Paris Regime. Glob Environ Polit 22:38–58
Jordan A, Huitema D, Hilden M, van Asselt H, Rayner T (2015) Emergence of polycentric climate governance and its future prospects
Krogstrup S, and Oman W (2019) Macroeconomic and financial policies for climate change mitigation; International Monetary Fund: Washington, DC, USA, vol 19
Kuyper JW, Schroeder H, Linnér B (2018) The evolution of the UNFCCC. Ann Rev Environ Resources
Ostrom E (1997) A behavioral approach to the rational choice theory of collective action. Am Pol Sci Rev 92:1–22

Ostrom E (2009) A polycentric approach for coping with climate change. World Bank Policy Research Working Paper Series, 5095
Sabel , C. and Victor D. (2017). Governing global problems under uncertainty: making bottom-up climate policy work. Climatic Change 144, 15–27.
Talbot Wright H, Buchuk D, Alvarado J, Vogt-Schilb A, Torres Pelaez D, Delgado R (2023) What can carbon prices achieve in Latin America and the Caribbean? Inter-American Development Bank
UNIDO (2022) The future of industrialisation in a post-pandemic world. Industrial development report 2022. UNIDO, Vienna
van Asselt H., and Zelli, F. (2018) International governance: polycentric governing by and beyond the UNFCCC. In: Jordan A, Huitema D, Van Asselt H, Forster J (eds) Governing climate change: polycentricity in action? Cambridge University Press, Cambridge, pp 29–46
Vandenbergh MP, Gilligan JM (2017) Beyond politics: the private governance response to climate change. Cambridge University Press
Victor DG, and Keohane RO (2010) The regime complex for climate change. Perspect Polit 9:7–23
Zelli, F. (2011). The fragmentation of the global climate governance architecture. WIREs Climate Change 2(2):141–308

Chapter 2
Carbon Pricing in Agriculture and Livestock

Elisa Belfiori and Daniel Perczyk

Abstract This chapter studies the role of the agricultural and livestock sector in climate change. This sector is a major source of greenhouse gas emissions in developing countries, but adopting sustainable farming and livestock practices these emissions could be reduced significantly. The chapter highlights the large potential for storing carbon in soils, discussing the role of carbon offsets in a global carbon market. The chapter argues that agriculture and livestock are key to addressing climate change in Argentina and that the development of carbon markets is essential for making the most of this opportunity of reducing emissions.

Keywords Carbon pricing · Agriculture · Livestock · Soil carbon sequestration · Climate policy

2.1 Introduction

The greenhouse gases accumulated in the atmosphere are the driver of anthropogenic global warming, and human activities are largely responsible for this, especially the burning of fossil fuels for electric generation, industrial, and transportation purposes. While fossil fuel combustion for energy and transportation remains the dominant source of emissions globally, in Latin America—and particularly in countries like Argentina—agriculture, livestock, and land use changes also play a significant role.

Argentina contributes approximately 1.7% of global emissions. At first glance, this may seem not very important. Even if the country reduced its emissions to zero, the global trend would not dramatically shift. But many countries have less than 2% of global emissions. The sum of the values of emissions from these countries represents more than 40% of global emissions. A legitimate question arises: Should Argentina

E. Belfiori (✉)
School of Business, Universidad Torcuato Di Tella, Buenos Aires, Argentina
e-mail: ebelfiori@utdt.edu

D. Perczyk
Fundación Torcuato Di Tella, Buenos Aires, Argentina
e-mail: dp@itdt.edu

E. Belfiori and G. Gertner (eds.), *Climate Change and Economic Growth*,
SpringerBriefs in Latin American Studies, https://doi.org/10.1007/978-3-031-99637-5_2

prioritize climate policy? The answer is unequivocally yes. Climate policy is increasingly taking the form of trade policy. Global markets are demanding low-carbon, sustainable products, and carbon adjustment tariffs are starting to emerge. Failing to meet low-carbon standards risks exclusion from key international markets—a scenario with significant economic consequences could result. Therefore, addressing climate change is an environmental challenge and a crucial economic and trade policy issue.

In the global context, agriculture, livestock, and land-use changes contribute to about 21% of global greenhouse gas emissions. However, for Argentina, this share is much higher—around 39%, nearly matching the contribution of the energy sector. Table 2.1 summarizes Argentina's emissions from the main categories in agriculture and livestock, compared to global emissions from this sector. Because Argentina's emissions are split roughly 50–50 between energy and land-based sectors, any effective national climate strategy must seriously engage with the agricultural sector. While the global climate agenda often prioritizes the energy sector, for Argentina, tackling emissions from agriculture and livestock is essential.

From a regional perspective, Latin America exhibits relatively higher emissions from agriculture, forestry, and other land uses, and lower emissions from industrial and energy processes. Emissions from the agricultural sector are significantly higher when measured on a per capita basis compared to Europe, North America, and the global average as Table 2.2 shows. This is in part driven by the lower population density that skews the per capita emissions upward.

Efforts to reduce emissions from agriculture are not only necessary but may also be more feasible in the short term than transforming the energy matrix. This is because many sustainable agricultural practices reduce emissions while also enhancing productivity and climate resilience. These practices include conservation agriculture, cover cropping, crop rotation, and improved livestock feeding techniques. Given its economic importance and technological sophistication and availability of natural resources, Argentina's agricultural sector is well-positioned to lead in this space.

2.2 Carbon Sequestration in Carbon Markets

From an economic perspective, the textbook solution to climate change is to price greenhouse gas emissions. Excessive emissions occur because market prices for goods and services do not account for the environmental cost of carbon. A price for carbon emissions aligns private incentives with social costs by making emitters pay for the external damage of carbon emissions.

The carbon price can take the form of a carbon tax or arise from an emissions trading system. Implementing a Pigouvian carbon tax would require emitters—from fossil fuels and agricultural activities—to pay for the environmental damage they cause. On the other hand, those who contribute to carbon capture should be rewarded

Table 2.1 Sectorial emissions: global and Argentina (2019)

Sector	Source	Global 2019	Argentina 2018	Global (%)	Argentina (%)
		(Mt CO_2eq)	(Mt CO_2eq)		
Crops	Synthetic fertilizer application	442.5	1.3	0.7	0.4
	Rice cultivation	1003	5.43	1.7	1.5
Subtotal crops		1445.5	6.8	–	–
Cattle	Enteric fermentation	2950	55.5	5.0	15.2
	Manure management	413	1.7	0.7	0.5
Subtotal cattle		3363	57.2	–	–
Land use, land-use change, and forestry	Managed soil and pasture	1475	41.3	2.5	11.3
	LULUCF	6490	– 2.0	11.0	– 0.6
Total LULUCF		7965	39.28	–	–
Total agriculture and LULUCF		12,773.5	143.2	21.7	39.1

Note LULUCF refers to Land use, Land-use change, and Forestry. Own elaboration based on data on world total anthropogenic direct and indirect GHG emissions for the year 2019 (in GtCO2eq) by sector and subsector (Fig. TS.6) in Climate Change 2022: Impacts, Adaptation, and Vulnerability. Contribution of Working Group II to the Sixth Assessment Report of the Intergovernmental Panel on Climate Change. Argentina data based on National Inventory of GHG.

for their efforts, creating a balanced system that incentivizes both emission reductions and carbon sequestration.

This is where carbon markets intersect with carbon sinks. Carbon markets can create new opportunities to finance carbon sequestration. While most current crediting mechanisms focus on reforestation and forest conservation, soil carbon sequestration may offer even greater potential. In a review of seven major carbon removal technologies, Griscom et al. (2017) estimate that soils could sequester up to 5 gigatons of CO annually by 2050—more than any other natural or technological solution, including forests.

Argentina's vast agricultural lands position the country uniquely to harness its potential in carbon sequestration. By focusing on soil carbon capture, Argentina can

Table 2.2 AFOLU sector—emissions per region

Region	% of Total sector GHG emissions	Per capita GHG emissions (t CO_2eq)
Australia, Japan, New Zealand	2	2
Africa	20	2
East Asia	9	1
Europe	3	0.8
Latin America	26	5.5
North America	4	1.8
South Asia	10	1
South East Asia	22	4.2
World	100	1.9

Source Agriculture, Forestry, and Other Land Uses (AFOLU). Own elaboration based on IPCC, 2022: Climate Change 2022: Mitigation of Climate Change. Contribution of Working Group III to the Sixth Assessment Report of the Intergovernmental Panel on Climate Change, Fig. 2.21

play an important role in global climate mitigation efforts while leveraging its natural resources.

Efforts have been made to promote using no-tillage to access carbon credits. No-tillage is an innovative and widely adopted agricultural system used by approximately 85% of Argentina's farmland. This practice involves planting crops without disturbing the soil and leaving the residue of previous crops as ground cover. Its implementation has drastically reduced the risks of water and wind erosion, which were significant challenges some decades ago. No-tillage shifted the focus from managing soil deterioration to actively enhancing soil quality.

Importantly, no-tillage has a substantial impact on carbon emissions. Reducing soil disturbance prevents the release of carbon stored in the soil's organic matter into the atmosphere. Additionally, it decreases fuel consumption by minimizing mechanical soil preparation. These combined effects result in carbon accumulation in the soil.

Current frameworks often require GHG reduction and capture projects to demonstrate that the projects wouldn't be implemented without the income related with the carbon credits, a principle known as additionality.

The emphasis on additionality can be a feature of voluntary carbon markets, where no regulated carbon price exists. In the absence of a market price on carbon emissions entering or leaving the atmosphere, voluntary markets assume that positive emissions will continue unabated and only reward additional sequestration efforts.

The distinction of whether or not an additionality criterion must be met is particularly important when considering nature-based solutions for carbon capture. Technological methods for carbon removal inherently meet the additionality requirement, as such investments would not occur without a financial incentive.

As comprehensive carbon markets develop, recognizing the value of all carbon sequestration, including that provided by existing natural systems, is essential for building a fair and efficient climate policy framework.

2.3 Measurement Is the Challenge

Given Argentina's vast and productive agricultural lands, there is an urgent need to improve the country's capacity to measure and verify soil carbon sequestration. This requires the development of national systems to quantify net emissions from agriculture—accounting for both emissions and carbon removals through sustainable land use. Strengthening this capacity will enable the agricultural sector to fully participate in emerging carbon markets and benefit from carbon credits.

Beyond environmental benefits, no-tillage enhances agricultural productivity. Emission metrics should account for emissions per unit of food produced. Higher productivity enables farmers to transform carbon into food more effectively. Recent innovations include the integration of cover crops, which maintain green fields year-round. These crops continue to photosynthesize and absorb carbon dioxide, further bolstering the environmental benefits of no-tillage systems.

Argentina already demonstrates leadership in low-emissions agriculture, not only with the widespread adoption of no-tillage practices but also with efficient logistics that result in lower agricultural emissions from food production compared to countries like the USA and China.

Table 2.3 shows Argentina's carbon footprint in its main agricultural and livestock products, compared to these countries.

Advantages such as direct sowing and the proximity of agricultural zones to ports contribute to lower fuel consumption for logistics, enhancing Argentina's competitiveness in sustainable farming. Agricultural production has achieved high efficiency in the production of milk, poultry, and pork.

On the other hand, cattle production is predominantly extensive, and its carbon footprint is slightly higher than that of more intensive systems in other countries. At the same time, grazing systems in Argentina are located in landscapes with exceptional capacity for carbon sequestration and biodiversity conservation—advantages that are unique at a global level.

Despite these advancements, significant challenges remain. Issues such as deforestation and the need for reforestation incentives extend beyond the fertile Pampas region to more marginal areas. Nitrogen fertilizers pose challenges, contributing to nitrous oxide emissions and evaporation losses. Addressing these issues will require changes in fertilizer types and application methods to minimize environmental impacts.

Livestock farming adds complexity due to methane emissions, a major contributor to agricultural carbon emissions. However, these emissions are often measured without considering the carbon capture from pastures, grasslands, and feed crops that are part of livestock systems. To move forward with sustainable practices, it

Table 2.3 Emissions by product in different countries

Product	Country	Unit	Value	Reference
Corn	China	kg CO_2 eq/ha	4052	(Zhang et al. 2017)
		kg CO_2 eq/kg corn	0.48	(Zhang et al. 2017)
	USA	kg CO_2 eq/kg corn	(0.36–0.44)	(Adom et al. 2012)
	Argentina	kg CO_2eq/ha	1247.8	(Maizar 2023)
		kg CO_2eq/kg corn	0.178	(Maizar 2023)
Wheat	China	kg CO_2 eq/ha	5455	(Zhang et al. 2017)
		kg CO_2 eq/kg wheat	0.75	(Zhang et al. 2017)
	USA	kg CO_2 eq/kg wheat	(0.38–0.51)	(Adom et al. 2012)
	Argentina	kg CO_2eq/ha	532.8	(Bongiovanni et al. 2023)
		kg CO_2eq/kg wheat	0.186	(Bongiovanni et al. 2023)
Soybean	China	kg CO_2 eq/kg soybean	(0.20–0.39)	(Zhang et al. 2023)
	USA	kg CO_2 eq/kg soybean	(0.33–0.52)	(Adom et al. 2012)
	Argentina	kg CO_2eq/kg soybean	0.150	(Viglizzo 2017)
Rice	China	kg CO_2 eq/ha	11,881	(Zhang et al. 2017)
		kg CO_2 eq/kg rice	1.6	(Zhang et al. 2017)
	Argentina	kg CO_2eq/kg rice	0.804	(INTA Informa 2022)
Cattle	USA	kg CO_2 eq/t beef	17.94	(Capper 2009)
	Argentina	kg CO_2eq/t LW	37	(Arrieta et al. 2022a, b)
		kg CO_2eq/kg HcW	53.4	(Arrieta et al. 2022a, b)
Dairy	USA	kg CO_2 eq/kg FPCM	(0.69–1.45)	(Rotz et al. 2021)
	New Zealand	kg CO_2 eq/kg FPCM	(0.773–0.806)	(Ledgard et al. 2020)
	Argentina	kg CO_2eq/l milk	1.09	(Arrieta et al. 2022a, b)

Sources Own elaboration

will be crucial to develop systematic methods for measuring both emissions and the carbon capture provided by agricultural lands. This will help Argentina strengthen its approach to sustainability in agriculture and livestock farming.

The Third National Greenhouse Gas Inventory included a special edition on mitigation scenarios. According to the report, there is a total potential of 2 million hectares for forestry projects, with an estimated mitigation potential of 55 million tons of CO equivalent per year (0.055 Gt CO_2eq/year). Other studies in the academic literature estimate carbon sequestration in soils to range between 180 and 205 CO_2eq/ha/year Alvarez et al. (2021). However, the total potential for carbon sequestration in soils in Argentina remains uncertain as researchers have not yet agreed on the potential per hectare or the total area that could be effectively targeted (Table 2.4).

This assessment concludes the likely range of global land-based mitigation potential is approximately 8–14 $GtCO_2$-eq yr^{-1} between 2020 and 2050 with carbon prices up to USD100 tCO_2-eq–1, about half of the technical potential (medium evidence,

Table 2.4 Mitigation global potential estimates for different agricultural options under various carbon price scenarios

Mitigation option	Estimate type	< USD20 $GtCO_2eq$-1	< USD50 $GtCO_2eq$-1	< USD100 $GtCO_2eq$-1	Technical
Agriculture total	Sectoral	0.9 (0.5–1.4)	1.6 (1–2.4)	4.1 (1.7–6.7)	11.2 (1.6–28.5)
	IAM	0.9 (0–3.1)	1.3 (0–3.2)	1.8 (0.7–3.3)	ND
Agriculture—carbon Sequestration (soil carbon management in croplands and grasslands, agroforestry, and biochar)	Sectoral	0.5 (0.4–0.6)	1.2 (0.9–1.6)	3.4 (1.4–5.5)	9.5 (1.1–25.3)
	IAM	ND	ND	ND	ND
Agriculture—reduce CH_4 and N_2O emissions (improve enteric fermentation, manure management, nutrient management, and rice cultivation)	Sectoral	0.4 (0.1–0.8)	0.4 (0.1–0.8)	0.6 (0.3–1.3)	1.7 (0.5–3.2)
	IAM	0.9 (0–3.1)	1.3 (0–3.2)	1.8 (0.7–3.3)	ND

Source Estimated annual mitigation potential ($GtCO_2$-eq yr^{-1}) in 2020–2050 of AFOLU mitigation options by carbon price. IPCC AR6 https://www.ipcc.ch/report/ar6/wg3/downloads/report/IPCC_AR6_WGIII_Chapter07.pdf

medium agreement). About 30–50% could be achieved if carbon price < USD20 tCO_2-eq.

2.4 Concluding Remarks

Soil carbon sequestration offers a major mitigation opportunity for the region, with the capacity to turn into a key player. The key challenge is the creation of robust institutional frameworks to support these efforts. Effective systems to measure, report, and verify net agricultural emissions are essential for accessing both voluntary and regulated carbon markets. While international agreements have traditionally focused on reforestation, soil carbon sequestration is an equally important mitigation pathway. The growing voluntary carbon markets, driven by private sector initiatives, offer Argentina new opportunities to participate and lead.

Research to better assess the carbon capture potential of soils and sustainable farming practices must be a priority. In parallel, building the necessary institutional structures will allow Argentina to integrate its agricultural sector into carbon crediting mechanisms, positioning the country as a leader in sustainable agriculture and climate action.

In sum, Argentina—and Latin America as a whole—has both the responsibility and the opportunity to lead in agricultural carbon markets. Investing in institutional frameworks, data systems, and policy tools to integrate agriculture into carbon pricing mechanisms is essential. This is not only a climate imperative but also a strategic economic opportunity for the region.

Acknowledgments We thank Pablo Cañada for his valuable assistance with the preparation of tables and data for this chapter.

References

Adom F, Maes A, Workman C et al. (2012) Regional carbon footprint analysis of dairy feeds for milk production in the USA. Int J Life Cycle Assess 17:520–534. https://doi.org/10.1007/s11367-012-0386-y

Alvarez R, Berhongaray G, Gimenez A (2021) Are grassland soils of the pampas sequestering carbon? Sci Total Environ 763:142978

Arrieta E, Cuchietti A, Cabrol D, González A (2018) Greenhouse gas emissions and energy efficiencies for soybeans and maize cultivated in different agronomic zones: a case study of Argentina. Sci Total Environ 625:199–208

Arrieta EM, Aguiar S, González Fischer C, Cuchietti A, Cabrol DA, González AD, Jobb´agy EG (2022) Environmental footprints of meat, milk and egg production in argentina. J Clean Prod 347:131325

Arrieta EM, Aguiar S, González Fischer C, Cuchietti A, Cabrol DA, González AD, Jobbágy EG (2022) Environmental footprints of meat, milk and egg production in Argentina. J Clean Prod 347:131325. https://doi.org/10.1016/j.jclepro.2022.131325

Bongiovanni R, Tuninetti L, Espósito G, Fiant S, Ponieman K, Álvarez S, Oliverio N (2023). LA HUELLA DE CARBONO DEL MAÍZ COMO VÍNCULO ENTRE PRODUCCIÓN Y AMBIENTE. Ciencia Del Suelo 41(2):157–179

Capper JL (2009) The environmental impact of beef production in the United States : 1977 compared with 2007 4249–4261. https://doi.org/10.2527/jas.2010-3784

de Moraes Sá JC, Lal R, Cerri CC, Lorenz K, Hungria M, de Faccio Carvalho PC (2017) Low-carbon agriculture in south America to mitigate global climate change and advance food security. Environ Int 98:102–112

Griscom BW, Adams J, Ellis PW, Houghton RA, Lomax G, Miteva DA, Schlesinger WH, Shoch D, Siikam¨aki JV, Smith P, Woodbury P, Zganjar C, Blackman A, Campari J, Conant RT, Delgado C, Elias P, Gopalakrishna T, Hamsik MR, Herrero M, Kiesecker J, Landis E, Laestadius L, Leavitt SM, Minnemeyer S, Polasky S, Potapov P, Putz FE, Sanderman J, Silvius M, Wollenberg E, Fargione J (2017) Natural climate solutions. In: Proceedings of the national academy of sciences, vol 114, issue no. 44, pp 11645–11650

INTA (2022) Estiman la huella de carbono de la producción arrocera. https://intainforma.inta.gob.ar/estiman-la-huella-de-carbono-de-la-produccion-arrocera/

Ledgard SF, Falconer SJ, Abercrombie R, Philip G, Hill JP (2020) Temporal, spatial, and management variability in the carbon footprint of New Zealand milk. J Dairy Sci 103(1):1031–1046. https://doi.org/10.3168/jds.2019-17182

Maizar (2023) Huella de carbono del maíz de Argentina. 2021–2022. http://www.maizar.org.ar/documentos/trabajohuellacarbono2023.pdf

Rotz A, Stout R, Leytem A, Feyereisen G, Waldrip H, Thoma G, Holly M, Bjorneberg D, Baker J, Vadas P, Kleinman P (2021) Environmental assessment of United States dairy farms. J Clean Prod 315:128153. https://doi.org/10.1016/j.jclepro.2021.128153

Viglizzo E, Frank F (2014) Energy use in agriculture: Argentina compared with other countries, pp 77–98

Viglizzo EF (2017) La Huella de Carbono en la Agroindustria (Issue January 2014)
Zhang Q, Hong J, Zhang T, Tian X, Geng Y, Chen W (2023) Environmental footprints of soybean production in China. Environ Dev Sustain 25(9):9047–9065. https://doi.org/10.1007/s10668-022-02424-1
Zhang D, Shen J, Zhang F, Li Y, Zhang W (2017) Carbon footprint of grain production in China 1–11. https://doi.org/10.1038/s41598-017-04182-x

Part II
Energy Supply and Energy Transition

Chapter 3
The Political Economies of Energy in Latin America

Peter Newell and Hayley Stevenson

Abstract Transitioning to energy systems that ensure secure supplies, equitable distribution and environmental sustainability is one of the most important challenges facing countries in the twenty-first century. Different countries are inserted into the regional and global political economy in different ways, and these differences shape transition pathways. As a region, Latin America reflects the competing energy transitions unfolding across the world: countries committed to leaving fossil fuels in the ground sit alongside others whose economies remain highly dependent on these resources. The availability of alternative energy sources offers no guarantee of progress towards sustainable systems. Energy transitions are political processes, not merely technical ones. This chapter offers an account of the political economies of energy in Latin America before focusing on the case of Argentina. This case illustrates how even a country with tremendous renewable potential can remain dependent on fossil fuels. The experience of recent governments in Argentina reveals how energy policy is shaped by political and economic factors that make it difficult to resolve trade-offs between stability, equity and sustainability.

Keywords Energy transition · Latin America · Political economy · Energy trilemma · Argentina

3.1 Introduction

As a region, Latin America is inserted into an uneven global political economy of energy transitions. Trade and investment relations, production networks and development assistance and intricate political, diplomatic, economic and environmental

P. Newell (✉)
University of Sussex, Brighton and Hove, UK
e-mail: P.J.Newell@sussex.ac.uk

H. Stevenson
Universidad Torcuato Di Tella, Buenos Aires, Argentina
e-mail: hstevenson@utdt.edu

E. Belfiori and G. Gertner (eds.), *Climate Change and Economic Growth*,
SpringerBriefs in Latin American Studies, https://doi.org/10.1007/978-3-031-99637-5_3

ties mean that what happens in the energy sector in Latin America has global ramifications and vice versa. Energy is central to the growth strategies of all economies, but many Latin American countries are especially dependent on resource revenues and the size of major economies in the region means their energy policy choices resonate globally and geopolitically, just as their integration into global markets means they are vulnerable to shocks and shifts in energy economies in other parts of the world.

Despite the challenges of transitioning energy economies in the region, progress is being made. Indicators of energy efficiency have consistently increased over the past 22 years, with the rate of improvement in 2012 43% higher than it was in 1990. Yet, the region's annual rate of improvement still lags behind the world average (0.9% compared to 2.1%) (Urteaga and Hallack 2021). Latin America's performance is more impressive on the renewable energy front where it is a global leader. The proportion of renewables reaches nearly 33% of total energy consumption. Due to high dependence on hydropower, Latin America has lower carbon electricity compared to the rest of the world with over 61% of electricity generation coming from renewables (Quiroga Martínez 2023: 3). But there is significant potential to further develop renewable energy and move away from fossil fuels. Despite vast potential to tap into its solar and wind resources, especially in countries such as Argentina, Chile, Colombia and Mexico, these still account for only 16% of the total renewable generation according to the International Renewable Energy Agency (IRENA 2022).

There is no singular energy transition unfolding across the region, but rather multiple competing ones. Within Latin America, we can observe simultaneously bold moves to leave fossil fuels in the ground (in Ecuador, Colombia, Costa Rica and Chile) and move almost entirely away from fossil fuels (Uruguay, Costa Rica), a gas rush and renewed interest in nuclear energy (Argentina) and drives towards hydrogen and lithium booms as part of a 'green' economy (Chile). This sits alongside major oil-dependent economies like Venezuela that would consider an energy transition away from fossil fuels a threat to their economic welfare and others with both abundant fossil fuels and strong renewable capacity that are taking a bet each way such as Mexico or Brazil where Petrobras plans to allocate a substantial 80% of its capital expenditure over the next five years to oil and gas exploration despite Brazil having secured $2bn in sustainable bonds for green investments. This complex and varied picture is not unique to Latin America and to different degrees is reflected across the world where major leaders on renewable energy such as China and India also continue to be major producers and consumers of fossil fuels. Indeed, this is one of the problematic features of energy transitions: new sources add to the energy mix but are not currently displacing fossil fuel production (York 2012) highlighting the need for measures to reduce demand.

The global landscape which forms the backdrop to events unfolding in the region is shaped by a number of contemporary and historical developments which give rise to these competing pathways. Geopolitically, the rising power of countries like Brazil within and beyond the region, alongside that of China, India and South Africa is reconfiguring the energy landscape around both fossil fuels and renewable energy (Power et al 2016; Shen and Power 2017). Momentous disruptions like the war in

Ukraine have also sent shock waves through the international system with implications for Latin America as countries like Argentina seek to position themselves as an attractive location for investments in gas which might substitute for exports from Russia, while countries part of the lithium triangle (Chile, Bolivia and Argentina) seek to benefit from rising demand for lithium and other minerals essential for batteries that underpin the drive towards electrification and away from fossil fuels. Part of this demand, in turn, is driven by the adoption of Green New Deal policies in Europe and the USA in the form of the Inflation Reduction Act under the administration of President Biden. Hence, shifting patterns of global supply (of resources) and (consumer and business) demand sharply impinge upon energy pathways in Latin America.

From the UN climate regime and the environmental community, the pressure is to decarbonise economies in the context of a global 'emissions gap' between pledges contained in countries nationally determined contributions and what is required to deliver the ambition of the Paris Agreement to keep warming below 1.5° compared with pre-industrial levels. Compounding this is the 'production gap' between planned levels of fossil fuel extraction and the level compatible with the goals of the Paris Agreement (currently 110% more than is compatible with the 1.5 target) (SEI & UNEP 2021). This enhances concerns about countries, including in Latin America, being left with 'stranded assets' and unburnable carbon (McGlade and Ekins 2015). At the same time, it is important to recognise both the historically minimal contribution many countries in the region have made to the problem of human-induced global heating and that some of the greatest contributions to global efforts to achieve net zero might come from non-energy sectors like land use change measures, to protect the Amazonian rainforests for example or to reduce emissions from the livestock sector in Argentina. Nevertheless, as parties to the climate agreements, all countries in the region are expected to report their progress in decarbonising their economies.

Taken together, this global backdrop presents an energy trilemma for Latin American countries seeking to address energy security through affordable and accessible energy in a shifting, conflictual and uncertain global environment (described above), seeking to avoid carbon lock-in and stranded assets while also seeking to address issues of energy access and poverty which still affect many groups in the region. Hence although access to electricity is estimated to be around 96% across the region (including the Caribbean), these figures disguise racial, class, gender and regional differences in energy access between wealthier and poorer parts of countries, even within major cities and between metropolitan centres and rural areas. There are also a multitude of non-climate-related environmental (and health) challenges associated with the energy sector from climate change to air quality and biodiversity impacts (since many reserves lie under biodiversity hotspots such as the Yasuni reserve in Ecuador). Hence, energy politics in the region intersects with the politics of resources, land rights, the rights of indigenous communities and many other issues besides which further complicate the political economies of energy transition. It is also strongly shaped by broader processes of economic reform underway whether around power sector reform and privatisation (Grimson and Kessler 2005), the removal of subsidies for energy and fossil fuels, or the governance of state-owned enterprises such as PDVSA (Petróleos de Venezuela), Pemex in Mexico, Petrobras in Brazil or

Repsol YPF in Argentina. Positively, as state-owned entities, these companies have more autonomy to align energy investments with broader state goals around energy security and sustainability since they are less beholden to shareholder pressure to deliver short-term returns and more concerned with job creation and the achievement of national goals. That same insulation from some commercial imperatives can also mean, however, that they are slower to reform and less subject to international pressures to rapidly decarbonise.

3.2 Understanding the Political Economies of Energy

The political economy of energy in the region can be understood in a number of ways. Historically, colonial relations played a key part in constructing economies along extractivist lines as Uruguayan political ecologist Eduardo Gudynas (2015) has pointed out, drawing in turn on Galeano's famous work on the open veins of Latin America (1997). Open veins refer to the construction of infrastructures (such as ports, railways lines and pipelines) and trading routes to minimise costs between sites of extraction and getting them to markets in the colonial centres of the world economy. Political economists from the dependency school of development have also pointed to the uneven nature of relations between the core and periphery as evidence of the globally exploitative dynamics of capitalism (Cardoso and Faletto 1979). Scholarship on ecological economics would provide an ecological reading of these uneven patterns of energy and resource extraction and exchange (Hornberg 1998) that continue today, evident, for example, in the assemblages of extractivism that source raw materials from regions where land and labour are cheaper and export them to the core of the global economy for processing or consumption and underpinning electrification initiatives for decarbonising the global North (Soto Hernandez et al. 2022). These are the spatial fixes by which richer countries can outsource the costs and risks of transition to poorer ones while maintaining dominant accumulation strategies (Newell 2021).

In seeking to understand the nature of energy transitions now underway across the region, scholarship on socio-technical transitions would help to understand how, when and why energy systems change with reference to the role of niche, regime and landscape actors. Niches would be the renewable energy actors seeking to secure market share and state support to decarbonise the economy, while regime actors would be dominant incumbent configurations made up of key state ministries and major actors in the energy sector. The landscape refers to many of the pressures and factors described in the previous section. Such approaches, and particularly the multi-level perspective which explores the interaction of these three levels (Geels and Schot 2007), constitute the dominant way of understanding transitions. They help to account for some of the key drivers of energy transition which are taken to be finance, technology and innovation. Where they are weaker is in providing an adequate account of power, politics and governance which we suggest is crucial to understanding the current nature and future prospects of energy transitions in the region and globally. This is vital to comprehending incumbent resistance to energy transitions and openings for change and the adoption of alternative energy pathways (Newell 2021). Change requires new alliances, coalitions and disruptive politics as well as attention to cultural political economy: the norms, values and behaviours that enable and constrain shifts in energy cultures (Strauss et al 2013).

Alongside these general accounts of the politics and governance, scholarship from and about the region sheds light on the importance of deeper transitions in Latin America: from authoritarianism to democracy (O'Donnell et al 2013), different democratic traditions which also determine the scope for citizen engagement with energy transitions (O'Donnell 2007), the ways in which different institutional structures shape energy policy (Hochstetler 2021), the politics of clientelism and the intersection between energy and trade politics for example (Newell 2007), all of which help to form the basis of the peculiar political economy of energy in Latin America that we turn to next.

Understanding the Political Economies of Energy in Latin America

As noted above, Latin America is a diverse region in terms of the size of countries, their political economies and the nature of their energy sectors. There is a very mixed picture of 'gas politics' in Bolivia, Chile and Argentina with intra-regional dimensions and tensions and global implications. There are entrenched petro-politics in Venezuela, Ecuador and Colombia, albeit to some extent now challenged by moves towards supply-side policies in Colombia, Chile and Ecuador. In the case of Colombia, the administration of Petro in Colombia has committed to not issue licences for new oil in Colombia, Chile is phasing out coal, while Ecuador oversaw an historical national referendum in August 2023 on the future of the Yasuni reserve and whether to drill for oil, and voted to keep it in the ground. Costa Rica, as one of the original members of the Beyond Oil and Gas Alliance, has committed to not issue licences for new oil and gas. There is strong pushback in each of these countries against these ambitious policies, in the wake of a new government in Costa Rica, the Ecuadorian government threatening not to honour the popular vote on Yasuni and significant resistance in Colombia to Petro's proposals to end new oil licenses. The plan to stop oil exploration and banning fracking announced in 2022 is already coming under fine and looks set to be a key election issue in 2026. Commentators have observed, 'By 2023 it is clear that his vision is too contentious and too threatening to economic stability, and that compromises and reversals are certain' (Colombia Risk Analysis 2022). Alternative energy pathways exist, but they are currently under-supported and fossil fuel interests will not roll over lightly whatever the stated intentions of the new government (Newell et al. 2023a, b).

Energy politics in the region are heavily influenced by global political developments and demands with China financing hydro and nuclear and a key source of demand for biofuels and minerals; and global demand for batteries fuelling minerals extractivism, especially lithium in the lithium triangle (Argentina, Bolivia, Chile) driven by global drive for electrification (Soto Hernandez et al. 2022). Between them Chile, Argentina and Bolivia are home to 58% of the world's identified lithium reserves, but thus far have also been unable to develop battery manufacturing assets that would move them up the value chain, reinforcing the sorts of concerns raised by critical scholarship noted in the previous section about the reproduction of regional inequalities. Countries are also positioned differently in terms of their ability to attract private and public climate finance. Climate funds (mainly from the Green Climate Fund, the Clean Technology Fund, the Amazon Fund and the Global Environment Facility) in the region are highly concentrated in Brazil, Costa Rica and Colombia, with a large share going to mitigation, including forest protection and reforestation, while market-based approaches, especially carbon pricing, have been adopted in Argentina (carbon tax), Chile (carbon tax and ETS) and Brazil (Emissions Trading Scheme).

It is not as straightforward as to assume resource abundance determines the make-up of the energy sector. As we illustrate in more detail below, Argentina, for example, has vast under-used potential for wind and solar energy, but lack of access to capital and an unattractive investment environment have inhibited their development (Newell 2014). There are important political economy factors in terms of access to credit and capital markets, levels of taxation on imported equipment and investor concerns about the 'investment climate' in countries bankrolling state-owned enterprises such as Petrobras in Brazil and Repsol YPF in Argentina with billions in subsidies and in the case of Argentina high levels of both debt (with a $65 billion foreign debt restructuring) and inflation.

Each country is inserted differently into the regional and global political economy in ways which greatly influence their energy sectors and scope for their transition. For some, there are *opportunities* available from global moves to transition away from fossil fuels, from lithium production (Chile and Argentina) and green hydrogen (Chile and Peru), exporting hydropower and biofuels (Brazil), the generation of forestry offset projects (Brazil and many Central American countries) but also *threats* associated with restrictions on financing fossil fuels (affecting gas projects like Vaca Muerta in Argentina) and investor fears over stranded assets affecting demand in heavily fossil-fuel-dependent economies (such as Venezuela). Opportunities and threats are usually the product of slow-moving social, political and economic processes, but geopolitical shocks can generate rapid reconfigurations (such as Russia's invasion of Ukraine which made Venezuelan oil more attractive in 2022).

An important issue then is how much policy autonomy countries in the region will have to chart their own preferred energy pathways in the face of a mix of investor, donor, public and civil society pressure (depending on the country in question). Energy issues are notoriously sensitive in trade negotiations, but energy is a sector that has been targeted by companies bringing cases through investor state dispute settlements. Mexico's experience of the North American Free Trade Area (NAFTA) treaty and provisions under Chapter 11 whereby the government was sued in August 2000 by the California-based Metalclad corporation for $16.7 million for rejecting its proposal to build a hazardous waste facility in an already 'highly contaminated' community might be a cautionary tale (Newell 2007). The World Bank and IDB have been active in trying to support the decarbonisaton of energy and transport systems in the region, sometimes controversially around hydro development, for example, but elsewhere encouraging the uptake of renewable energy in rural areas such as the World Bank (US$30 million) and GEF (US$10 million) funded PERMER project in Argentina (Proyecto de Energías Renovables en Mercados Rurales) (PERMER 2009).

For transitions to be socially and politically acceptable in Latin America, they will have to be socially just. This means addressing issues of retraining and compensation for workers in sectors that need to transition (such as oil workers and coal miners), regional redevelopment to create alternative employment opportunities, as well as attending to an array of social inequalities (along gender, race, class lines). For example, about 90% of Colombian coal is extracted from the remote northern regions of La Guajira and Cesar, which are characterised by ethnic diversity and high levels of poverty (Strambo et al. 2018). In Cesar Colombia involuntary resettlement was ordered by the government's own Ministry of Environment and Sustainable Development (Reyes 2020). The resettlement plan sought to generate mechanisms of information, communication and participation to involve the (directly and indirectly) impacted population as well as the local and regional authorities in all stages of the participatory resettlement plan. This included reinstating by collective and individual agreement the productive activities of the affected communities in relation to land, crops and agrarian infrastructure (Reyes 2020).

It is not a coincidence that Argentina was one of the first countries to promote the idea of just transitions in the climate negotiations amid these concerns about the socially differentiated impacts of transition (Newell and Mulvaney 2013). There is scope for regional and global institutions to help support such a transition. For them to perform such a role, however, their approach would need to be more plural and inclusive than that observed to date through Just Energy Transition Partnerships and the overriding emphasis on de-risking private finance as opposed to broader issues of economic justice affecting the region such as high levels of indebtedness (Newell et al 2023a, b). In the absence of efforts to attend to these issues, countries across the region can expect resistance from communities whose prior and informed consent is often not secured, their lands acquired and promised benefits often not forthcoming. This can result in conflict and violence as has been observed around wind projects in Mexico (Dunlap 2018; Hesketh 2002; Torres Contreras 2023). Likewise in Colombia, concerns have been reported around the legitimate representation of indigenous communities during consultation processes for wind energy projects in La Guajira region, Colombia—where indigenous Wayúu represent half of the population. This includes issues around unbalanced company—community negotiations over compensation and undue pressure suffered by the community both from the company and from community advisors leading to potentially unfair agreements for the communities (Vega-Araújo and Heffron 2022). In terms of raising justice issues, indigenous movements have played a key role in resource politics in the region (Hesketh 2023). Most recently, they played a key role in the national referendum on the future of oil in the Yasuni reserve in Ecuador noted above, but have a long history of shaping and resisting energy projects and infrastructures in the region from Initiative for the Integration of the Regional Infrastructure of South America (IIRSA) to the resistance of the U'wa people in Colombia to Occidental's oil pipelines and waves of litigation brought by groups in Ecuador against Texaco (Collinson 1996; Garvey and Newell 2004). Indeed, Temper et al. (2020) found that over a quarter of fossil fuel projects encountering social resistance have been cancelled, suspended or delayed.

3.3 The Case of Argentina

Argentina's attempts to diversify its energy sources reveal how even a country with tremendous renewable potential can remain stuck in dependence on fossil fuels. This case also reveals the challenge of avoiding trade-offs among the three features of a healthy energy system: stability, sustainability and equity. Two energy transitions took place in Argentina in the twentieth century, and each consolidated the dominance of fossil fuels. The first, in the 1920s, saw Argentina begin to exploit domestic petroleum resources to reduce dependence on imported coal and foreign companies. The world's first state-owned petroleum company was created, *Yacimientos Petrolíferos Fiscales* (YPF) to promote the construction of infrastructure and the diffusion of 'oil nationalism' throughout the country (Sabbatella et al. 2020: 80–1). The second transition began in the 1940s with the Perón government seeking to expand domestic energy supplies as part of a broader strategy of reducing imports and promoting national industrialisation and self-sufficiency. This saw the creation in 1945 of another state-owned energy company, *Gas del Estado*, and investments in infrastructure for producing and distributing liquefied petroleum gas (LPG) and later compressed natural gas (CNG). This strategy succeeded in increasing the share of gas in Argentina's energy matrix, but domestic supplies could not keep up with growing demand so the country became increasingly dependent on foreign imports. This, in turn, generated interest in hydropower and later nuclear power. Additional state-owned energy companies were established in the 1960s (HIDRONOR S.A. and *Agua y Energía Eléctrica S.E*), and in partnership with private capital, they constructed considerable hydro infrastructure in various parts of the country (Radovich et al. 2012a, b). The first nuclear power plant, Atucha I, began operating in 1974, with a second coming online in 1984. By 1990, hydropower accounted for over 38% of electricity generated, nuclear 16% and fossil fuels 46% (Jensen et al. 2019).

But confidence in these alternative sources of energy was affected by the energy crisis of 1988–89 when residents in Buenos Aires (*porteños*) were forced to endure scheduled power cuts for 5–6 h each day during the hot summer months. The energy system's capacity to meet demand was already strained when, in 1988, Argentina suffered one of its worst droughts on record, which affected hydroelectricity generation (Red GIRCyT 2016: 30). Also in 1988, the Atucha I nuclear plant suffered a malfunction in its fuel channels, sparking fears of a Chernobyl-like disaster. The problem was resolved with local ingenuity and without the support of its German designers (whose services were too costly and reliant on unattainable parts), but the plant remained offline for eighteen months (Vitale 2023; Almagro et al. 2017). Ultimately, this incident injected some confidence into Argentina's capacity to produce nuclear power with local technology and know-how, and a third power plant was eventually completed in 2014, but the sector's contribution to the energy matrix never surpassed its 1990 peak.

The 1988/89 energy crisis coincided with the government of Carlos Menem's mass privatisation of productive industries and infrastructure. This new context favoured the expansion of fossil fuel infrastructure (especially natural gas-fired thermal power plants), which could be built faster and at a lower cost than hydropower dams and nuclear plants. With reduced state support for alternative energy, a 're-fossilisation process' began: the fossil fuel industry strengthened its historic dominance and the share of non-fossil fuels in the energy matrix gradually declined (Ruggeri and Garrido 2021).

The twenty-first century has seen renewed efforts to strengthen Argentina's energy system by increasing the share of renewable sources in the energy mix. Nevertheless, these efforts have failed to achieve their objectives and have often failed to adequately consider stability, sustainability and equity. Below we illustrate this with reference to the Kirchner and Macri governments' diversification strategies, as well as offer reflections on challenges faced by the government of Alberto Fernández and that will now have to be addressed by his successor Javier Milei elected in November 2023.

Between 2003 and 2015, governments led by Nestor Kirchner and then Cristina Fernández de Kirchner sought to reassert the state's control over energy development. In the wake of the 2001 financial crisis, the interim Peronist government resisted pressure from business by freezing gas rates in the local currency, thereby decoupling gas from international market prices. This helped stimulate economic activity, but the growth in energy demand coincided with a period of decline in oil and gas production (Ruggeri and Garrido 2021). The incoming government led by Nestor Kirchner responded by maintaining the rate freeze and subsidising utility companies to compensate for low tariffs, expanding energy imports to meet short-term demand and introducing mechanisms to encourage investment and expansion of domestic production capacity (Bril-Mascarenhas and Post 2014: 108). How effective were these strategies from the perspective of the energy trilemma?

According to official data (INDEC), the number of urban households living in conditions of 'energy poverty'[1] fell from 40% in 2003 to less than 1% in 2015 (Durán and Condorí 2019: 433). Comparable data for rural households is unavailable, but it is clear that the reduction in energy poverty was less impressive in rural areas because many are not connected to the electricity grid and missed out on consumer subsidies (Durán and Condorí 2016; Bril-Mascarenhas and Post 2014: 110–11). The distribution of subsidies was also regressive within urban areas because they were universal rather than targeted according to need. This meant that wealthier households benefited more because they consume more and have a higher rate of connection to the grid (Cont et al. 2011: 9–10). The drop in urban energy poverty was achieved at a tremendous—and unsustainable—cost to the state. By 2010, consumer subsidies for gas, electricity and transport had reached 10% of the national budget (US$7.5 billion) (Bril-Mascarenhas and Post 2014: 100). Consumers were paying for only 30% of

[1] Defined as households forced to spend more than 10% of their income to access sufficient energy.

their power bills with the state covering the rest (Godoy and Calcagno 2016). The subsidy-based approach also negatively impacted the stability dimension of healthy energy systems. Artificially low prices acted as a disincentive to investors, so while consumption grew, production declined and imports increased (Barrera 2021).

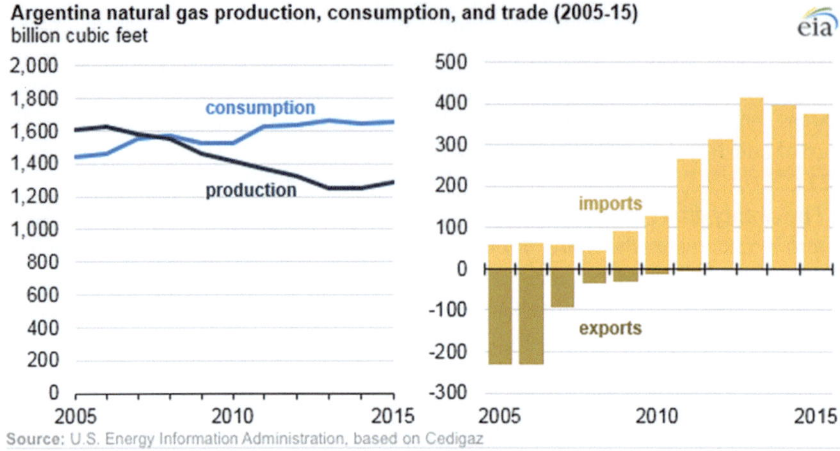

Between 2005 and 2015, total installed capacity increased by about 35% (EOL 2021) and much of the additional capacity was due to state investments financed largely through Chinese and Russian loans (Sabbatella et al. 2020: 83–4). Only a small proportion of this increased capacity came from non-fossil fuels (nuclear, hydropower and biofuels). The Kirchner government passed a law in 2006 declaring it a national interest to generate electricity from renewable sources and establishing a target of sourcing 8% of electricity from these sources by 2016. 'Small' hydro was included in the law (up to 50 MW), but larger dams would not count as renewable energy (Presidencia de la Nación 2006). Environmental concerns did not drive this law, but rather an interest in reducing dependence on foreign imports by expanding local capacity. The law introduced tax incentives to promote private sector investment in wind and solar power, small hydropower dams and biomass, but these incentives could not overcome the disincentive imposed by low energy prices (relative to international markets) (Sabbatella et al. 2020: 85–6). The government's response to this problem was the GenRen program, which allowed the state-owned company Energía Argentina S.A. (ENARSA) to purchase electricity generated by private companies using renewable energy. The program offered 15-year power purchase contracts at a fixed price in dollars in exchange for conditions that equipment be produced nationally (Ruggeri and Garrido 2021). However, GenRen failed to reach its objectives due to low levels of investor confidence in the state's guarantees combined with the global financial crisis and restrictions imposed on Argentina by international lenders, which made it difficult for local investors to access capital at reasonable rates (ibid; Sabbatella et al. 2020: 85–6). Despite the Kirchner governments' interest in promoting domestic production (especially of wind turbines), imported machinery was used because the local industry did not have the volume to supply even the limited number of completed wind power projects (Ruggeri and Garrido 2021).

By 2015, renewable sources still accounted for less than 2% of installed capacity (compared to 60.9% for fossil fuels and 32% for large hydro) (Ferreiro 2022). The renewable energy law was modified at this time to postpone the initial target to 2017, set additional targets leading up to 20% of electricity generation by 2025 and create a fund to lend capital to potential investors (Congreso de La Nación Argentina 2015). In this context, a new neoliberal government led by Mauricio Macri took power. In sum, energy policy under the Kirchner governments had mixed results from the perspective of equity, but quite poor results from the perspectives of stability and sustainability.

The neoliberal government of Mauricio Macri took power in 2015 and swiftly began to liberalise the energy system, re-dollarising tariffs, substantially reducing subsidies and deregulating capital markets. Between 2015 and 2019, electricity tariffs rose by between 3400 and 5500%, while gas tariffs rose on average between 1300 and 1500% (Giménez 2021). As a result, during this period nearly 3 million households fell into energy poverty or energy destitution (defined as spending more than 20% of income on energy) (UNLP 2021). Climate change concerns were articulated more explicitly by the Macri government, but efforts to promote renewable energy were motivated by a desire to re-insert Argentina into international markets and demonstrate commitment to the G20 club and the Paris Agreement on climate change. Efforts during these years do not reflect a transition to a healthy energy system; renewable sources were not considered as alternatives to fossil fuels, but as an additional technology for improving supplies, reflective of the tendency noted above (York 2012). Moreover, there was no effort to promote resilience in the energy system by considering the impacts of future climatic changes such as drought.

Macri maintained the fund created under the 2015 modified law and bolstered it with support from the World Bank; this added a guarantee for investors in the event of non-compliance by the state. Investor support was also facilitated by resolving the country's historic debt with foreign creditors, which saw Argentina welcomed back into international credit markets. Conditions tied to domestically produced equipment were eliminated and reduced import tariffs allowed investors to take full advantage of international markets in which technology prices had fallen considerably (Ruggeri and Garrido 2021). This generated opposition from local producers (represented by the industry group, CITERA: Cluster of Industries and Technologies of Renewable Energies of Argentina), as well as from trade unions and the Peronist political movement who opposed dollarised tariff increases (labelled '*tarifazos*'). But opposition groups had limited impact, in part because many members of the powerful oil industry bought into the program rather than obstructing it. Several of the renewable projects built during the Macri years were led by oil companies, including Pan American Energy/BP, Pampa Energía and YPF (Barrera et al. 2022), in transition terms illustrative of how incumbent actors can seek to control market access for new 'niche' actors.

The Macri government's market-based approach to expanding renewable energy capacity proved effective in attracting investors: 198 renewable energy projects were signed (mostly for wind and solar) and by 2021 the share of renewable sources had grown to 11%. While more impressive than earlier growth rates, it fell short of expectations and missed the interim targets set in 2016 (which required 12% by 2019). From a sustainability perspective, it is notable that '44% of the growth in installed power generation capacity between 2015 and June 2021 was explained by renewable energy' (Barrera et al. 2022). One factor explaining the underperformance of Macri's renewable energy program is the domestic economic crisis: in its early years, the vast majority of signed projects secured full financing and became operational, but in later years fewer and fewer projects were completed (Barrera et al. 2022). Another factor is the insufficient attention given to state responsibilities. The Macri government inherited legislation that makes the state responsible for energy transport infrastructure. They tried to maintain a market orientation by using public–private partnerships, but this approach was unsuccessful, and the transmission network remained stagnant as capacity of both thermal and renewable energy grew (Ruggeri and Garrido 2021).

In sum, energy policy under the Macri government had very poor results from the perspective of equity, and mixed (albeit improved) results from the perspectives of stability and sustainability. While not directly related to energy policy, this government's decision to take out a $45 billion loan from the IMF will inevitably affect Argentina's energy system for many years to come. Repaying the debt requires foreign currency and exporting commodities is perceived as the most viable way to obtain such currency. In this context, the widely accepted, but highly questionable, discourse of gas as a 'transition fuel' is being embraced across the political spectrum in Argentina. The shale gas project of Vaca Muerta is consistently mentioned as key to Argentina's energy transition, with the International Energy Agency forecasting a record increase to 1 million barrels per day by 2028. But underinvestment in infrastructure to distribute energy also affects the gas sector. Vestas, the largest renewable energy company operating in Argentina, cites such underinvestment as the major weakness in Argentina's energy system (Jamele 2023). Despite the fact that macroeconomic instability makes it hard to operate in Argentina, Vestas has been able to complete multiple projects, but the grid has all but collapsed and cannot take anymore. In this context, Vestas insists that investing in distribution infrastructure is more important for now than increasing installed capacity (ibid). By the end of 2022, renewable energy accounted for less than 14% of generated electricity while thermal sources still accounted for 59% (Ministerio de Economía 2023). It is hard to see how the 2025 target of 20% will be met, but in 2023 the Fernández government expanded this target to 57% by 2030, calculating that it will cost over $86 billion to achieve (Koop 2023).

These are the challenges that the government of Javier Milei now inherits and despite his intention to liberalise large swathes of the economy, including the energy sector and the company YPF, he will likely face fierce resistance from government civil servants, labour unions and other social movements opposed to such moves. It is too early to be able to differentiate between populist bluster and actionable reforms that he will undertake. But given his professed climate scepticism and proximity to business interests likely to be opposed to more ambitious action on the environment, the prospects for greater leadership from Argentina appear quite dim. As Colombo puts it, 'environment is one of the ministries that will cease to exist with the inauguration on 10 December since the newly elected far-right president has promised to reduce his ministerial cabinet from 18 to eight portfolios—eliminating those that 'are not a priority'. A climate change denier, Milei has promised not to comply with the 2030 agenda—which is centred on 17 sustainable development goals—and to withdraw from the Paris Agreement because he will not accept 'impositions from outside'. He says his government will not have policies to fight the climate crisis, protect Indigenous people, or decrease deforestation, nor to sustainably regulate the production of the shale oil and gas reserve of Vaca Muerta, seen in Argentina as a golden goose' (Colombo 2023). For the sake of Argentina and the rest of the world, we have to hope the rhetoric is not translated into reality.

3.4 Openings and Ways Forward

If an energy transition is to take place across Latin America at the speed and scale required to deliver on ambitious climate targets, it will require widespread social acceptance (including of some of inevitable costs and disruption that attend any such transition) and international support with economic diversification, technological capacity-building and retraining workers in key sectors of the economy. Financing, not just through traditional climate and carbon finance, but potentially also addressing debt relief will be required. Latin American countries were at the forefront of experiments with debt-for-nature swaps in the 1990s, and compensation mechanisms were explored again at the time of the Yasuni-ITT initiative (Sovacool and Scarpaci 2016) such that Ecuador would have received a percentage of the expected value of the oil that would otherwise have been extracted. In the wake of the national referendum in August 2023 to keep the oil in the ground, the debate about financial support to the country has been reignited and has implications for other countries in the region.

Questions of historical responsibility for climate change (low for most of the region) and capacity to transition will need to be assessed on a country by country basis. But further action will almost certainly be conditional on more ambitious action forthcoming from the richer 'core' of the global economy and those highly industrialised countries historically and contemporarily most responsible for (re)producing the problem. As the UNFCCC puts it, responsibility is 'common but differentiated'. A more persuasive and engaging narrative for energy transition might be around climate resilient or climate compatible development (Nunan 2017), simultaneously recognising key development needs, the region's vulnerability to the effects of climate change (on agriculture, infrastructures and health for example) and the need to reduce emissions from the region and avoid lock-in to higher carbon pathways which will leave it isolated in the global economy of the future. This might include strong emphasis on co-benefits, enabling conditions for poorer groups which prioritise systems of provision which benefit the poor but avoid high carbon lock-in (low carbon affordable housing; affordable public transport, paid for by congestion charges or taxes on high emitting vehicles; mini and micro-grids, community energy for rural areas off grid), and means of enhancing employment and improving health through air quality and thereby making huge savings on health expenditure. As Latin American unemployment reaches the highest level in a decade, and 45 million people are forced into poverty, just transitions will have to be based on a combination of creating new job opportunities for workers in the energy sector of the future while protecting indigenous rights and encroachments on their land and livelihoods from expanded fossil fuel frontiers or as sacrifice zones for renewable energy generation. To try and reduce the need for ever expanding supply, governments will need to also support reduced demand through better urban planning around housing, transport and localised 'smart' grids.

Latin America stands at a cross-roads. With many countries in the region such as Costa Rica, Chile, Colombia and Ecuador demonstrating climate leadership in different ways, energy transitions represent an opportunity to build lower carbon and more resilient energy systems of the future. But the pathway to pursuing such a vision, in particular how inclusive and socially just it is, will determine how successful it will ultimately be.

References

Almagro JC, Perazzo R, Sidelnik JI (2017) Crónica de una reparación (im)posible: el incidente de 1988 de la C.N. Atucha I. Ciudad Autónoma de Buenos Aires : Asociación Argentina para el Progreso de las Ciencias. Available at https://aargentinapciencias.org/wp-content/uploads/2018/01/LibrosDigitales/Cronica-de-una-reparacion-im-posible-Libro.pdf. Accessed 17 Oct 2023

Barrera MA (2021) El complejo energético argentino y los impactos estructurales sobre el sector externo. Ensayos De Economía 31:89–111

Barrera MA, Sabbatella I, Serrani E (2022) Macroeconomic barriers to energy transition in peripheral countries: the case of Argentina. Energy Policy 168:113117

Bril-Mascarenhas T, Post AE (2014) Policy traps: consumer subsidies in post-crisis Argentina. Stud Contemp Int Develop 50:98–120

Cardoso FH, Faletto E (1979) Dependency and development in Latin America. University of California Press, Berkeley

Collinson H (ed) (1996) Green Guerrillas: environmental conflicts and initiatives in Latin America and the Caribbean. Latin American Bureau, London

Colombia Risk Analysis (2022) 'What to Expect in the First 100 Days of Gustavo Petro's Government', August 16. https://www.colombiariskanalysis.com/post/what-to-expect-in-the-first-100-days-ofgustavo-petro-s-government. Accessed 17 Oct 2023

Colombo S (2023) 'From horrible to merely bad': will Javier Milei take his chainsaw to the environment in Argentina? The Guardian, 9th December https://www.theguardian.com/global-development/2023/dec/09/from-horrible-to-merely-bad-will-javier-milei-take-his-chainsaw-to-the-environment-in-argentina

Congreso de La Nación Argentina (2015) Ley 26190. Régimen de Fomento Nacional para el uso de Fuentes Renovables de Energía destinada a la Producción de Energía Eléctrica. Modificación, 21 de octubre. https://www.argentina.gob.ar/normativa/nacional/253626/texto. Accessed 17 Oct 2023

Cont W, Hancevic P, Navajas F (2011) Energy populism and household welfare. Energy Econ 56:464–474

Dunlap A (2018) Counterinsurgency for Wind Energy: The Bíi Hioxo Wind Park in Juchitán, Mexico. J Peasant Stud 3:630–652
Durán R, Condorí M (2016) Índice Multidimensional de Pobreza Energético Para Argentina: Su Definicion, Evaluacion y Resultados a Nivel de Departamentos para el año 2010. Acta de la XXXIX Reunión de Trabajo de la Asociación Argentina de Energías Renovables y Medio Ambiente, Vol 4, pp 12.27–12.38. Available at https://notablesdelaciencia.conicet.gov.ar/bitstream/handle/11336/85254/CONICET_Digital_Nro.683e7244-2066-4910-a937-a1102cc25d84_A.pdf?%20sequence=2&isAllowed=y. Accessed 17 Oct 2023
Durán R, Condorí M (2019) Evolución de la pobreza energética en Argentina durante el período 2002–2018. Oportunidades para las energías renovables. Extensionismo, Innovación y Transferencia Tecnológica: claves para el desarrollo 5:430–437
EIA (US Energy Information Administration) (2017) Argentina seeking increased natural gas production from shale resources to reduce imports, 19 February. https://www.eia.gov/todayinenergy/detail.php?id=29912. Accessed 17 Oct 2023
EOL (Energía Online) (2021) Generación de energía y potencia instalada en Argentina, 22 junio. https://energiaonline.com.ar/generacion-de-energia-y-potencia-instalada-en-argentina/. Accessed 17 Oct 2023
Ferreiro I (2022) Energías renovables: la Argentina aún está lejos de cumplir con los compromisos internacionales asumidos, 20 de febrero. https://chequeado.com/el-explicador/energias-renovables-la-argentina-aun-esta-lejos-de-cumplir-con-los-compromisos-internacionales-asumidos/. Accessed 17 Oct 2023
Galeano E (1997) Open veins of Latin America: five centuries of the pillage of a continent, anniversary. Monthly Review Press, New York
Garvey N, Newell P (2004) Corporate accountability to the poor? Assessing the effectiveness of community-based strategies. Dev Pract 15:389–404
Geels FW (2007) Schot J (2007) Typology of sociotechnical transition pathways. Res Policy 36:399–417
Giménez J (2021) Kicillof, sobre las tarifas en el gobierno de Macri: "Aumentaban 1300%, hasta 2000% el gas y 3700% la electricidad", 27 de agosto. https://chequeado.com/ultimas-noticias/kicillof-sobre-las-tarifas-en-el-gobierno-de-macri-aumentaban-1300-hasta-2000-el-gas-y-3700-la-electricidad/. Accessed 17 Oct 2023
Godoy F, Calcagno C (2016) New government set to introduce changes to energy sector, Beretta Godoy, 11 January. https://berettagodoy.com/wp-content/uploads/2016/01/208_New_government_set_to_introduce_changes_to_energy_sector_-_Newsletters_-_International_Law_Office.pdf. Accessed 17 Oct 2023
Grimson A, Kessler G (2005) On Argentina and the Southern Cone: Neo-liberalism and national imaginations. Routledge, London
Gudynas E (2015) Extractivismos: Ecología, Economía y Política de Un Modo de Entender El Desarrollo y La Naturaleza. Primera edición. CEDIB, Centro de Documentación e Información Bolivia, Cochabamba
Hesketh C (2002) Clean development or the development of dispossession? The political economy of wind parks in Southern Mexico. Environ Plann e: Nature and Space 5:543–565
Hesketh C (2023) Indigenous resistance at the frontiers of accumulation: challenging the coloniality of space in international relations. Rev Int Stud. https://doi.org/10.1017/S0260210523000268
Hochstetler K (2021) Political economies of energy transition: wind and solar in Brazil and South Africa. Cambridge University Press, Cambridge
Hornborg A (1998) Towards an ecological theory of unequal exchange: articulating world system theory. Ecol Econ Ecol Econ 25:127–136
IRENA (2022) Renewable capacity statistics 2021. https://www.irena.org/-/media/Files/IRENA/Agency/Publication/2021/Apr/IRENA_RE_Capacity_Statistics_2021.pdf. Accessed 17 Oct 2023

Jamele A (2023) Andrés Gismondi, de Vestas: “El problema no está en la generación de energía sino en el transporte y distribución”, Forbes 3 de marzo. https://www.forbesargentina.com/innovacion/andres-gismondi-vestas-el-problema-esta-generacion-energia-sino-transporte-distribucion-n30342. Accessed 17 Oct 2023
Jensen S, Zamora A, Rimancus P (2019) Evolución de la Matriz Eléctrica Argentina, 4 enero. https://estrucplan.com.ar/evolucion-de-la-matriz-electrica-argentina/. Accessed 17 Oct 2023
Koop F (2023) Argentina quiere lograr una gran expansión de las energías renovables para 2030, 20 de julio. https://dialogochino.net/es/clima-y-energia-es/374748-argentina-quiere-lograr-una-gran-expansion-de-las-energias-renovables-para-2030/. Accessed 17 Oct 2023
McGlade C, Ekins P (2015) The geographical distribution of fossil fuels unused when limiting global warming to 2 °C. Nature 517(7533):187–190
Ministerio de Economía (2023) Energías Renovables 2022: se cubrió el 13,9% de la demanda y se inauguraron 8 proyectos por más de 47 MW de potencia instalada, 31 de enero. https://www.argentina.gob.ar/noticias/energias-renovables-2022-se-cubrio-el-139-de-la-demanda-y-se-inauguraron-8-proyectos-por. Accessed 17 Oct 2023
Newell P (2007) Trade and environmental justice in Latin America. New Polit Econ 12:237–259
Newell P (2014) The politics and political economy of the CDM in Argentina. Environ Polit 23:321–338
Newell P, Mulvaney D (2013) The political economy of the just transition. Geogr J 179:132–140
Newell P (2021) Power shift: the global political economy of energy transitions. CUP, Cambridge
Newell P, Price R, Daley F (2023) Landscapes of (In)justice: reflecting on voices, spaces, and alliances for just transition, IDS Working Paper 594, Brighton: Institute of Development Studies. https://doi.org/10.19088/IDS.2023.047
Newell P, Daley F, Mikheeva O, Peša I (2023b) Mind the gap: The global governance of just transitions. Global Pol 14(3):425–437
Nunan F (ed) (2017) Making climate compatible development happen Oxon: Routledge
O’Donnell G (2007) Dissonances: democratic critiques of democracy. University of Notre Dame Press, Notre Dame
O’Donnell G, Schmitter PC, Whitehead L (2013) Transitions from authoritarian rule: Latin America. JHU Press, Baltimore
PERMER (2009) Proyecto de Energías Renovables en Mercados Rurales. Available from: http://energia.mecon.gov.ar/permer/intro_obj.htm. Accessed 17 Oct 2023
Power M, Newell P, Baker L, Bulkeley H, Kirshner J, Smith A (2016) The political economy of energy transitions in Mozambique and South Africa: the role of the rising powers. Energy Res Soc Sci 17:10–19. https://doi.org/10.1016/j.erss.2016.03.007
Presidencia de la Nación (2006) Boletín Oficial de la República Argentina, Poder Legislativo, Ley 26190/2006. 27 diciembre. https://www.argentina.gob.ar/sites/default/files/ley_26190-2006.pdf. Accessed 17 Oct 2023
Quiroga Martínez R (2023) Energy transition powering transformative sustainable development in Latin America and the Caribbean, Comisión Económica para América Latina y el Caribe. https://www.cepal.org/sites/default/files/news/files/ppt_challenges_eclac.pdf
Radovich JC, Balazote A, Piccinini D (2012a) Desarrollo de represas hidroeléctricas en la Argentina de la posconvertibilidad. Avá, (21) Recuperado en 18 de septiembre de 2025, de https://www.scielo.org.ar/scielo.php?script=sci_arttext&pid=S1851-16942012000200002&lng=es&tlng=es
Radovich JC, Balazote A, Piccinini D (2012b) Desarrollo de Represas Hidroeléctricas en La Argentina de la Posconvertibilidad, Avá 21. https://www.ava.unam.edu.ar/images/21/pdf/n21a02.pdf. Accessed 17 Oct 2023
Red GIRCyT (2016) Probabilidad de Perturbaciones Graves en el Abastecimiento Eléctrico por Amenazas de Origen Natural. https://www.argentina.gob.ar/sites/default/files/probabilidad_de_perturbaciones_graves_en_el_abastecimiento_electrico_por_amenazas_de_origen_natural.pdf. Accessed 17 Oct 2023

Reyes G (2020) Reasentamiento de el Hatillo, Cesar: La asimetría en la concentración entre comunidades y empresas mineras (P. Newell, Trans.). Bogotá: Pensamiento y Acción Social. https://www.pas.org.co/investigacion-el-hatillo

Ruggeri E, Garrido S (2021) More renewable power, same old problems? Scope and limitations of renewable energy programs in Argentina. Energy Res Soc Sci 79:102161

Sabbatella I, Barrera MA, Serrani, E (2020) Paradigmas Energéticos en Disputa en las Últimas Dos Decadas de la Argentina. In: Guzowski C, Martín MI, Zabaloy MF (eds) ENERGÍA, INNOVACIÓN Y AMBIENTE PARA UNA TRANSICIÓN ENERGÉTICA SUSTENTABLE. Editorial de la Universidad Nacional del Sur, Bahía Blanca, pp 79–94

SEI, IISD, ODI, E3G, UNEP (2021) The production gap report 2021.https://www.unep.org/resources/report/production-gap-report-2021

Shen W, Power M (2017) Africa and the export of China's clean energy revolution'. Third World Quart 38(3):678–697

Soto Hernandez D, Newell P, Blanco O (2022) Assembling extractivism in the lithium triangle. J Peasant Stud 49:945–968

Sovacool BK, Scarpaci J (2016) Energy justice and the contested petroleum politics of stranded assets: policy insights from the Yasuní-ITT initiative in Ecuador. Energy Policy 95:158–171

Strambo C, Espinosa ACG, Velasco AJP, Atteridge A (2018) Privileged coal: the politics of subsidies for coal production in Colombia, SEI Working Paper No. 2018–01, Stockholm Environment Institute (SEI) https://www.sei.org/wp-content/uploads/2018/03/sei-2018-pubs-coal-subsidies-political-0129.pdf. Accessed 17 Oct 2023

Strauss S, Rupp S, Love T (eds) (2013) Cultures of energy: power, practices. Left Coast Press, Walnut Creek, CA, Technology

Temper L et al (2020) Movements shaping climate futures: a systematic mapping of protests against fossil fuel and low-carbon energy projects. Environ Res Lett 15(12):123004

Torres Contreras GA (2023) Who owns the land owns the wind? Land and citizenship in the Isthmus of Tehuantepec, Mexico. J Agrarian Change 23:365–384

UNLP (Universidad Nacional de La Plata) (2021) Novedades: Bajó la Pobreza Energética durante el 2020. https://unlp.edu.ar/extension/medios/observatoriosp/novedades-bajo-la-pobreza-energetica-durante-el-2020-20467-25467/. Accessed 17 Oct 2023

Urteaga JA, Hallack M (2021) Eficiencia energética en América Latina y el Caribe ¿por qué importa? https://blogs.iadb.org/energia/es/eficiencia-energetica-en-america-latina-y-el-caribe-por-que-importa/

Vega-Araújo J, Heffron RJ (2022) Achieving energy justice in project development in La Guajira, Colombia. Energy Res Soc Sci 91:102688

Vitale S (2023) Atucha: el peor incidente de su historia y la asombrosa forma en que se solucionó, 17 mayo. https://www.lanacion.com.ar/la-nacion-revista/atucha-el-peor-incidente-de-su-historia-y-la-asombrosa-forma-en-que-se-evito-un-desastre-nid17052023/. Accessed 17 Oct 2023

York R (2012) Do alternative energy sources displace fossil fuels? Nat Clim Chang 2:441–443

Chapter 4
Challenges and Opportunities in Developing a Sustainable Energy Transition Strategy for Argentina

Walter Cont and Pedro I. Hancevic

Abstract Argentina, like most nations, faces the dual challenge of meeting its growing energy demands while simultaneously mitigating the adverse effects of climate change. As the global community increasingly turns towards sustainable energy sources, Argentina must develop a comprehensive strategy for transitioning its energy sector towards a more sustainable future. This chapter starts by examining the current situation and then moves towards the challenges and opportunities involved in developing a transition strategy, considering the country's unique economic context, natural resource endowment, and policy landscape. Drawing upon existing literature, case studies, and some qualified expert insights, the chapter provides recommendations for policymakers and stakeholders to navigate the complexities of the energy transition in Argentina.

4.1 Introduction

The urgent need for a transition to sustainable energy sources has been dramatically emphasized by the increasingly negative consequences of global warming and climate change. The urgency of this transition is further intensified when considering the serious threat posed by the climate crisis and the inherent geopolitical volatility affecting global energy security. Although previous attempts have been taken since the Kyoto Protocol, the formal starting point is the Paris Agreement, in force since 2015, which sought the commitment of countries to keep the increase in the average

W. Cont (✉)
CAF—Banco de Desarrollo de América Latina y el Caribe, Buenos Aires, Argentina
e-mail: wcont@caf.com

P. I. Hancevic
Centro de Investigación y Docencia Económicas (CIDE), Sede Región Centro, Aguascalientes, México

Universidad Panamericana Campus Aguascalientes, Aguascalientes, México

P. I. Hancevic
e-mail: phancevic@up.edu.mx

E. Belfiori and G. Gertner (eds.), *Climate Change and Economic Growth*,
SpringerBriefs in Latin American Studies, https://doi.org/10.1007/978-3-031-99637-5_4

global temperature below 2 °C (and preferably below 1.5 °C) compared to pre-industrial levels. Speculation has arisen about a possible and desirable scenario of net-zero CO_2 emissions by 2050, with partial targets in 2030, to meet the goals agreed upon in Paris (Clarke et al., 2014; Audoly et al. 2018; DeAngelo et al. 2021; Bataille et al. 2016; Rockström et al., 2017). It was estimated that to stay below 1.5 °C of global warming, emissions must be reduced by approximately 50% by 2030. However, experts from the United Nations Intergovernmental Panel on Climate Change (IPCC) have concluded, after assessing the individual commitments of each country, that these efforts will not be sufficient to achieve the agreed-upon temperature increase limits (Tong et al., 2019).

The commitments agreed upon by the parties at the COP 25 are voluntary proposals, albeit in a binding nature. From a theoretical point of view, a better option would have been the application of a global carbon tax, which would require a level of international coordination that only could be achieved through supranational control and enforcement. However, even the most fervent defenders of this alternative recognize its impracticality, given the prevailing geopolitical conditions (Cramton et al. 2020).[1]

In this context, most countries, in compliance with the Paris Agreement, have individually set targets for the reduction of greenhouse gas (GHG) emissions.[2] As complementary goals, and even considering them as mere policy instruments, several countries have set targets for the penetration (also called standards) of renewable energy, which usually require that a percentage of the energy produced or consumed come from clean renewable sources. In general, this percentage tends to increase over time, implying increasingly demanding clean energy standards. Likewise, some countries have established complementary programs such as those for energy efficiency, distributed generation and forced interconnection to the electricity grid, incentives for the development of biofuels, among other measures, through which they seek to help in the task of reducing GHG emissions.[3]

Argentina is not oblivious to this phenomenon on a global scale. The Law on Minimum Budgets for Climate Change Adaptation and Mitigation (Law No. 27,520 of 2019) provided a legal framework for policies and actions related to the reduction of GHG emissions and adaptation to the impacts of climate change. The Renewable

---

[1] As Ritz (2022) rightly points out, product markets are characterized by firm heterogeneity, market power, and the vagaries of international trade. Thus, the optimal policy of a hypothetical global social planner is second-best. The main idea is that optimal carbon prices can be very asymmetric: zero in some countries and higher than the social cost of carbon in countries with relatively dirty production. This result is obtained even though a uniform global carbon price always succeeds in reducing emissions from countries. Competition policy that mitigates market power can enable strong climate action.

[2] It goes without saying that the Paris Agreement is not limited to the energy sector, but covers all economic activities, including mining, agriculture, livestock, soil management, the forest industry, and among others.

[3] A source for tracking evolution at country level is the Climate Action Tracker website (climateactiontracker.org/).

Energy Law (Law No. 27,191 of 2015) established specific objectives for the incorporation of renewable energy sources and aimed to promote investments in sustainable projects. The National Climate Change Adaptation and Mitigation Plans of 2019 and 2022 and the Long-Term Resilient Low-Emission Development Strategy for 2050 proposed that the country must undergo an energy transition, focusing efforts on promoting energy efficiency, renewable energy, and distributed generation. During this period, natural gas would be considered as the main transitional fuel. The plan encompasses various sectors of the economy, such as energy, transportation, agriculture, and forestry, aiming to coordinate and enhance the country's actions to address the challenges of climate change. In addition to specific laws, Argentina has implemented sectoral policies to address climate change in areas such as sustainable agriculture, forest management, energy efficiency, and sustainable transportation.

Nevertheless, the Argentina's commitments and targeted goals derived from the Paris Agreement are not ambitious, falling short in the energy transition. The country presented its first NDC in 2016, a review in 2020, and a slightly more demanding revision at the Latin American Summit on Climate Change in 2021. Specifically, the Argentine government set a maximum limit of GHG emissions at 349 MtCO2 by the year 2030, implying a reduction of slightly more than 21% compared to its historical peak reached in 2007,[4] which is highly insufficient to meet the 1,5° domestic pathways, as shown in Fig. 4.1. In this chapter, we depart from these unambitious environmental goals and undertake an exercise more in line with climate urgencies, proposing a scenario of complete decarbonization of the economy and also reviewing the necessary transition for this process.

Interestingly, the energy industry was responsible for 78% of global GHG emissions.[5] In the Latin American and the Caribbean (LAC) region, this share significantly reduces to 47%, with the remaining 53% attributed to agriculture, forestry, and other land utilization (Cont et al. 2022). In Argentina, the National Greenhouse Gas Inventory (INGEI) reveals that a qualitatively similar situation as in LAC: about 51% of total emissions are energy related.[6] When disaggregating by subsectors, emissions are primarily attributed to transportation, accounting for 27.5% of total energy sector emissions, followed closely by electricity generation, contributing 21.7%. The subsequent categories responsible for energy source emissions include the use of industrial fuels (17.8%), residential fuel consumption (13%), fuel manufacturing (9.7%), fugitive emissions (5.9%), and, lastly, fuels from other sectors (4.3%).

[4] The emissions of Argentina in 2007, as reported in the Biennial Update Report (IBA, for its acronym in Spanish), were 441.4 MtCO2e. When converted using the PCG 4AR instead of the 2AR to facilitate comparability with the global emissions of the Emissions Gap Report (EGR), they amounted to 455.8 MtCO2e.

[5] Emissions Trends and Drivers Supplementary Material. En IPCC: Climate Change 2022: Mitigation of Climate Change. Contribution of Working Group III to the Sixth Assessment Report of the Intergovernmental Panel of Climate Change.

[6] National Greenhouse Gas Inventory and Monitoring of Mitigation Measures. Available at: https://inventariogei.ambiente.gob.ar/.

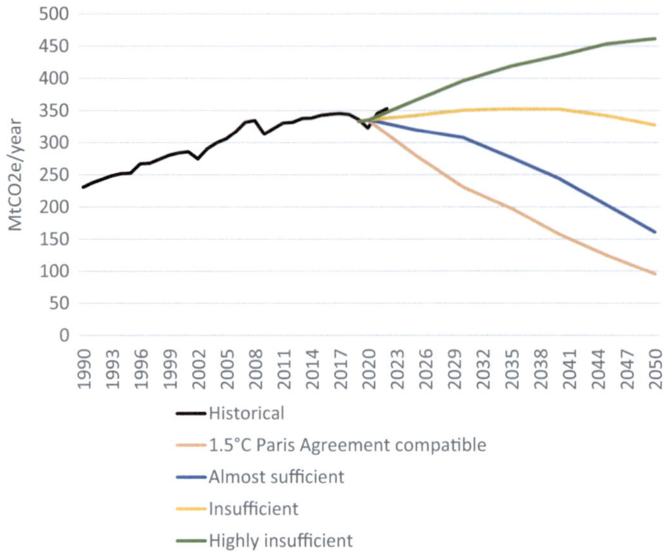

Fig. 4.1 Argentina overall rating regarding 1.5° domestic pathways. *Source* Own elaboration using data from Climate Action Tracker

In this chapter, we analyze a decarbonization scenario for Argentina that involves two fundamental changes to the status quo.[7] Firstly, achieving a net-zero emissions goal necessitates the almost complete abandonment of fossil fuel consumption. Secondly, there must be massive electrification across all economic sectors, starting with transportation and industry—the top two sectors in terms of fossil fuel consumption—followed by households, the commercial sector, and primary activities. In this context, the electricity generation meeting this enormous demand must come from clean renewable sources. Since the transition from one state to another will not occur automatically, we enumerate a set of policies that have been addressed for Argentina and briefly analyze them. Amid the massive electrification process, fuels such as (green and blue) hydrogen will emerge in certain industries and the transportation sector. Additionally, given the abundance of natural gas resources—particularly shale gas—and the high dependence of power generation and industry (as a heat source) on natural gas, we consider this fuel to be used during the energy transition. In the same line, but with enormous differences, biofuels will be important during the transition. Figure 4.2 schematizes a roadmap towards decarbonization in Argentina.

The rest of the chapter is organized as follows. Section 4.2 introduces the context for understanding the current situation in Argentina. Section 4.3 discusses the

[7] For a general framework and a global exercise of this type see, e.g., Bataille (2020). See also a roadmap proposed by Deloitte–Enel (2022) and the Guidelines and Scenarios proposed by the Undersecretary of Energy Planning (2023).

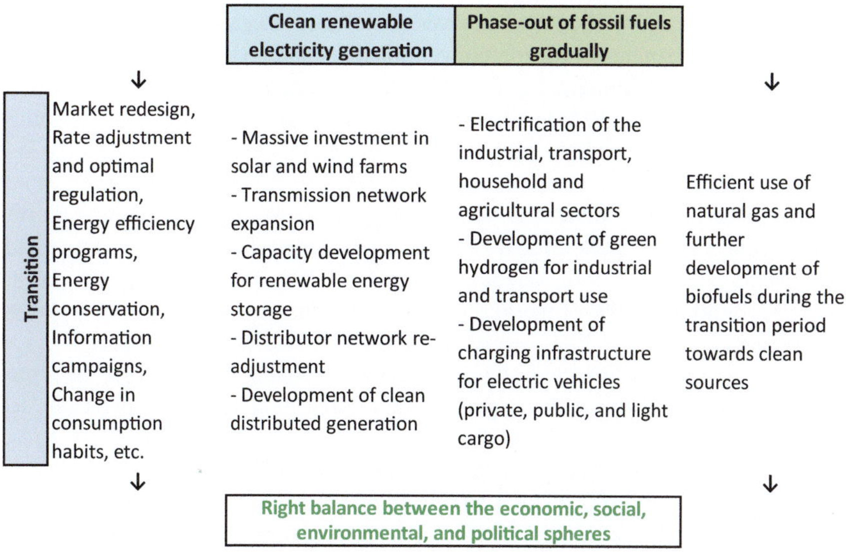

Fig. 4.2 Scheme of the Argentine energy transition. *Source* Adapted from Hancevic et al. (2023)

roadmap along three pillars: green generation, consumption substitution and efficiency, and complementary measures. Sections 4.4 and 4.5 discuss the challenges and opportunities for Argentina to pursue the net-zero path. Finally, Sect. 4.6 concludes.

4.2 The Context

In 2022, the country had an estimated population of 46 million, a GDP of 632 billion USD (in current dollars), with the overall energy sector representing more than 6% of the GDP.[8] Based on the 2022 National Energy Balance, there exists a substantial reliance on fossil fuels, comprising 87% of the total primary energy supply, equivalent to 3372 petajoules (PJ). Natural gas takes the lead contributing 54% to the overall domestic energy supply, followed by oil with a 37% share, while non-conventional renewable energy holds a 6% share (including wind and solar generation, and biofuels). Hydropower, nuclear power and coal each account for 2%–3% collectively complete the composition of the primary energy mix. Globally, Argentina holds the second-largest reserve of shale gas and the fourth-largest reserve of shale oil.[9] Current proved reserves of natural gas could meet 8 years of consumption, which would increase to 1619 years when including non-conventional

[8] The energy sector is understood to encompass oil and hydrocarbon exploitation, as well as the production of electricity, gas, and coal. Source: World Economic Outlook Database (2022) from IMF. Available at https://www.imf.org/en/Publications/WEO/weo-database/2022/April.

[9] Source: International Energy Agency (IEA). Available at: https://www.iea.org/countries/argentina.

resources. In the case of oil, proved reserves could meet 18 years of consumption, and up to 155 years when including non-conventional resources (IEA 2023, and OLADE 2023). In 2023, Argentina produced 132 MMm3 of natural gas and exported only 4 MMm3 (it reached over 30 MMm3 during the first decade of the twenty-first century to later close exports by 2007–2008). On the other hand, the country produced 635 k barrels per day (bpd) of oil and exported over 100 k bpd, but it still maintains a net importer status for oil products. In addition, Argentina stands as the world's fourth-largest lithium producer, a critical mineral for batteries and storage systems and, consequently, essential for the ongoing energy transition. Together with Bolivia and Chile, they hold over 60% of world reserves. Argentina could reach a production of 700,000 tons per year by 2040, representing about 20% of world production.

When attention is directed towards secondary energy sources, it becomes imperative to scrutinize the developments pertaining to electricity generation (see Fig. 4.4).

Clearly, thermal generation plays a major role in electricity generation in Argentina, averaging 60% of total supply (generation plus imports) over the last two decades. For instance, in 2022, 53% was from natural gas, and the remaining 7% came from liquid fuels and coal. Generation from renewable sources came from hydroelectric (17.5%), nuclear (5.4%), wind (10.3%), solar (2.1%), and biofuels (just 0.9%). Imports represented 4.6% that year. This sets a contrast with what occurs in other countries in Latin America and the Caribbean, where hydroelectric generation is predominant (e.g., Brazil, Paraguay, Ecuador, Costa Rica, and Colombia),

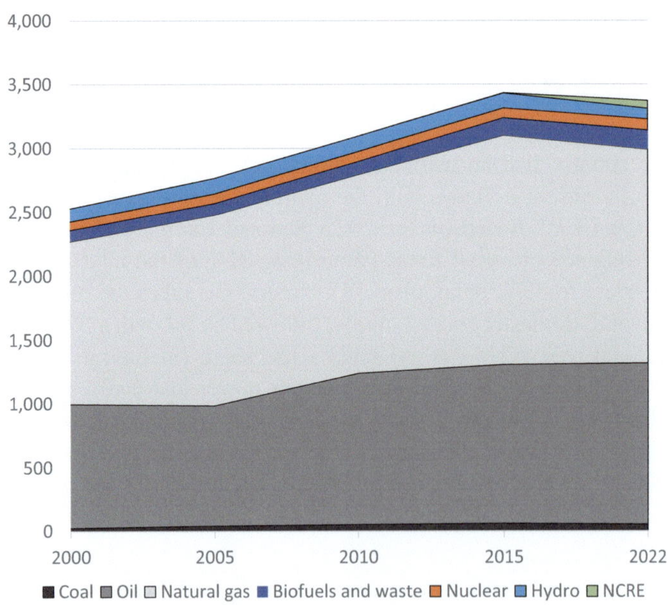

Fig. 4.3 Total energy supply in Argentina (in petajoules), selected years between 2000 and 2022. *Source* Own elaboration based on data from International Energy Agency (IEA)

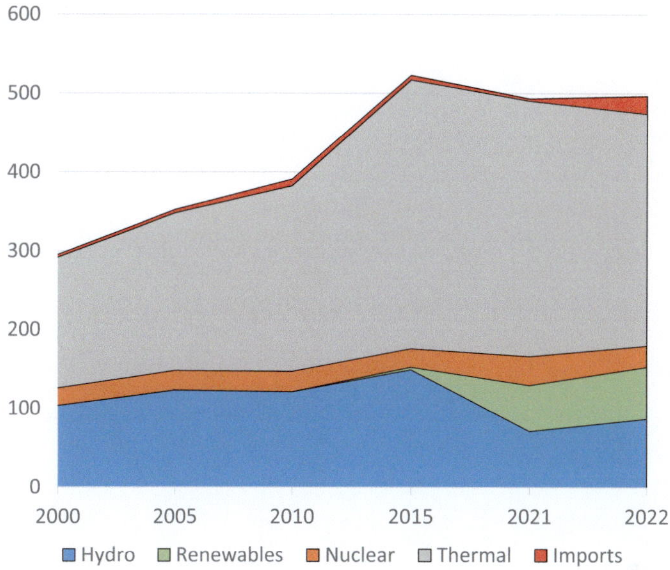

Fig. 4.4 Electricity generation by technology in petajoules (PJ). *Source* Own elaboration based on data from OLADE and CAMMESA Argentina

or where unconventional renewable sources dominate (as in Uruguay). The Argentinean case is closer to that of the Mexican electricity generation mix, where the dependence on imported natural gas from the USA is very significant (above 60%). It is worth highlighting, however, the rapid emergence of unconventional renewable energies in recent years, stemming from the implementation of the RenovAr (Argentina's Renewable Energy Auction) and MATER (Renewable Energy Forward Market) programs. These programs gained momentum between 2016 and 2019, but considerably slowed down since 2020.

Besides the aforementioned reserves of oil and natural gas, Argentina also possesses copious natural resources in the realm of renewable sources. Consequently, contemplating an entirely green scenario is by no means implausible.

Figure 4.5 presents solar and wind resources that make the country an incredible place to develop renewables. The wind power potential of Argentina exceeds 2000 GW (gigawatts), which is 65 times the total installed capacity in the country, considering all sources. Similarly, more than half of Argentina's territory receives annual average sunlight over 3.5 kWh/m^2. Notwithstanding solar and wind potential areas are geographically concentrated, in the northwest and south of Argentina, respectively. In addition, the gross theoretical hydro potential of Argentina is substantial and has been estimated in 169,000 GWh/year with a technically feasible potential of 130,000 GWh/year. But the green scenario faces another challenge ahead, for electricity consumption represents approximately 20% of total energy consumption.

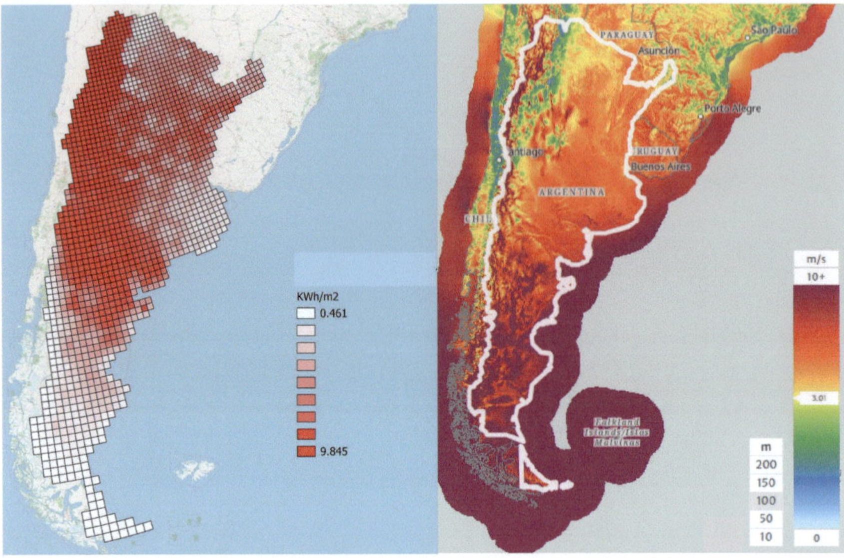

(a) Solar resources (normal indirect irradiance in kWh/m2) (b) Wind speed (m/s at 100 metres)

Fig. 4.5 Solar and wind resources in Argentina. *Source* Own elaboration using data from Ministry of Energy and Solar Wind Atlas

4.3 Pillars of the Energy Transition

4.3.1 Electric Power Generation Through Clean Energy Sources

The electricity sector is poised to become the primary driver of the energy transition in Argentina and probably in most economies of the world (Hancevic et al. 2023; Williams et al. 2012). In that sense, it is necessary to make a distinction between the path that conventional clean energies (mainly hydro and nuclear) may take and unconventional clean energies (mainly solar and wind).[10] In the first case, as mentioned previously, Argentina has a long tradition and high potential in hydroelectric generation, especially on a large scale. However, the RenovAr program has

[10] Marine energies, such as tidal and wave energy, represent a renewable resource with significant growth potential but still very unexplored in Argentina. The Argentine maritime platform, covering an area of 2,809,237 km^2 and boasting 5 thousand kilometers of coastline, is particularly conducive to the development of these energies. At the moment, there are no substantive projects in this area. The National Program for Research and Productive Innovation in Argentine Maritime Spaces (PROMAR), created by Law 27.167 in 2015, aims to promote the development of this sector. It would be desirable to make more efforts in that direction, combining public and private resources.

managed to incorporate small hydroelectric developments, which involve the generation of up to 50 MW of energy.[11] Regarding nuclear energy, Argentina is one of the few countries worldwide that masters the technology in all its stages. Unlike renewable energy sources such as hydroelectric, wind, or solar, nuclear generation is not directly dependent on climatic factors. On average, the annual contribution to electricity generation over the last fifteen years has been 5%, remaining very stable with minimal variation. Given the favorable conditions in the country for the development of this technology, it is desirable to propose and evaluate new projects that expand clean generation capacity through nuclear energy.

Now we move on to analyze unconventional clean energies, which undoubtedly should be the focus of attention for the energy transition. As illustrated in Fig. 4.5, the geographical and climatic conditions of each region will significantly influence the distinct development of various technologies. The southern region of Patagonia, for instance, is expected to predominantly harness wind power, while the northeast region is anticipated to contribute significantly to solar energy generation. In the case of water resources, 85% of surface water corresponds to the La Plata River basin,[12] but additional resources are distributed across the northern, central, and southern zones of the continental territory.[13] Currently, the country utilizes only one-fifth of its water resources for electricity generation, indicating that there is still immense potential for the development of new hydroelectric dams.

The key aspect is to be able to build enough clean electricity generation capacity to cover the entirety of the current electricity demand, as well as the additional demand that will be added as more sectors of the economy, particularly transportation and industry, are electrified. Under this challenging scenario, reliable infrastructure will be necessary. In particular, the expansion of the electricity transmission and distribution networks will be crucial to meet peak demand, provide energy reliability, and avoid disruptions in service. Figure 4.6 illustrates the evolution of high-tension and distribution lines in kilometers alongside the peak demand in MW for the corresponding time period. It is evident that the demand trend has not been met with a commensurate increase in transmission lines, especially since 2012. On this matter, the Secretary of Energy (2022) anticipates challenges on both the length of transmission lines and specific grid bottlenecks to accommodate geographic-specific non-renewable generation.

In the long run, a cleaner power grid will likely result in significant savings for consumers, as renewable generation costs are already competitive and are expected to become much lower than fossil fuel generation in the coming years. Figure 4.7 illustrates the levelized cost of energy (LCOE) in Argentina by energy source up to the year 2020. A fair comparison between technologies must encompass additional

---

[11] Argentina has a total of 75 micro and mini hydroelectric plants, with an installed capacity of 377 MW and a contribution of energy exceeding 1% of the national demand.

[12] This includes the sub-basins of the Pilcomayo, Bermejo, Paraná, Paraguay, Iguazú, and Uruguay rivers, and other sub-basins extending into the territories of Bolivia, Brazil, Paraguay, and Uruguay.

[13] Argentina also shares water resources with neighboring countries, as seen in the case of the Paraná River, where the binational Yacyretá dam is located.

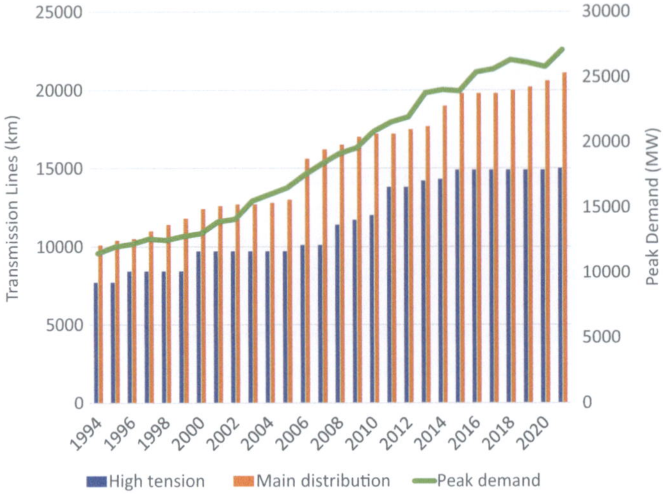

Fig. 4.6 Transmission lines and electricity demand evolution. *Source* Own elaboration based on data from CAMMESA

distribution and transmission costs along with considerations of carbon emissions and pollution avoidance resulting from the transition from fossil fuel-based generation to renewable-based generation.[14] However, even when including these elements, renewables are thought to be the most cost-effective source of electricity from 2035 onwards (Wood Mackenzie 2024).

4.3.1.1 Energy Storage and Imbalance Markets

Grid-scale storage will assume a crucial role in any net-zero emissions scenario. It will deliver vital system services that span short-term balancing and operating reserves, ancillary services to ensure grid stability, deferment of investments in new transmission and distribution lines, and long-term energy storage for the restoration of grid operations after blackouts. In this context, batteries will play a crucial role in mitigating the impact on the power grid by managing hourly and seasonal variations in renewable electricity output and ensuring grid stability and reliability amid the escalating demand. The cost of large energy storage systems has been consistently decreasing, and most analysts consider them to be an economically viable option. However, the expansion of electric vehicle production, disruptions in the market and competition from electric vehicle manufacturers have resulted in increased costs for essential minerals used in battery production, particularly lithium. It is increasingly

[14] Transmission costs will be higher in the case of a grid dominated by renewable energies due to the greater distances between consumption and generation centers as well as to the location specificity of the latter.

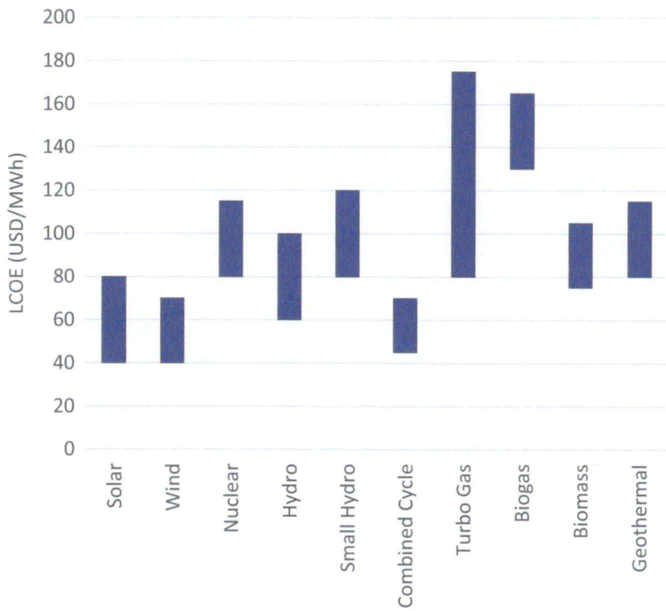

Fig. 4.7 Levelized cost of electricity (LCOE) in Argentina. *Source* Own elaboration based on data from the Ministry of Energy

apparent that achieving additional cost reductions hinges not only on technological innovation but also on the pricing dynamics of battery minerals.[15]

A complementary option to investments in utility-scale storage systems for newly added renewable generation and investments in transmission network expansion is the creation of imbalance markets (De Vries and Verzijlbergh 2018; Egerer et al., 2015; Ortner and Totschnig 2019). This option should be seriously considered for a regional scope that includes other countries in the region. The central idea is to have a platform that balances supply and demand fluctuations, automatically finding the lowest-cost electricity to meet real-time consumption. This way, congestion in transmission lines can be better managed, maintaining network reliability and supporting the integration of renewable energies. A key fact is that transmission investments tend to take much longer than generation investments. Therefore, it urges to study different possibilities for integrating multiple countries to take advantage of different availabilities of natural resources and to plan an integrated electric transmission and energy storage infrastructure. Argentina is already connected through high-tension power lines with Brazil and Chile (and through lower-tension power lines with Bolivia). Also, there are studies underway to improve electricity exchanges between the countries of the Southern Cone, through the existing interconnection infrastructure (SIESUR project).

[15] For more information, see https://www.iea.org/energy-system/electricity/grid-scale-storage.

4.3.1.2 Market Mechanisms for the Development of Renewables

In this section, we first provide a short description of each market mechanism and then analyze the Argentinean experience and provide some insights for future measures.

Market mechanisms for the development of renewable energy are designed to incentivize investment and growth in renewable energy projects. In addition to establishing government goals for transitioning to clean energy, market mechanisms are required to facilitate the adoption of renewable energies. Some of the most common mechanisms include:

- *Renewable Energy Certificates (RECs)*[16]: RECs allow producers to sell certificates representing renewable energy generation, providing an additional revenue stream. Buyers can support renewable energy even if they don't directly consume it, promoting demand for green power.
- *Feed-in Tariffs (FiTs)*: Governments or utilities offer long-term contracts with fixed payments to renewable energy producers for the electricity they generate. This guarantees a stable return and encourages investment by reducing market risk.
- *Renewable Portfolio Standards (RPS)*: These policies require utilities to source a certain percentage of their electricity from renewable sources. This creates a market demand for renewables, pushing utilities to invest in or purchase renewable energy to meet the standard.
- *Auction Systems*: Governments hold competitive bidding processes where renewable energy developers propose projects and the lowest bid wins. This system encourages cost efficiency while ensuring new renewable capacity is built, such as Argentina's RenovAr program, Mexico's initial REC auctions, Chilean or Brazilian mechanisms.
- *Carbon Pricing (Cap-and-Trade Systems or Carbon Taxes)*: By placing a price on carbon emissions, either through a cap-and-trade system or a carbon tax, these mechanisms make fossil fuels more expensive, making renewable energy comparatively cheaper and more attractive.
- *Subsidies and Tax Credits*: Governments may provide financial incentives like subsidies, tax breaks, or rebates for renewable energy projects. In many countries, tax credits (such as the U.S. Investment Tax Credit) have been critical in driving renewable energy growth.

Each of these mechanisms plays a role in encouraging investment, lowering the cost of renewable energy, and facilitating the transition away from fossil fuels. It is important to separate carbon taxes and cap-and-trade schemes from the rest as they are the only ones that target GHG emissions directly and therefore are more effective to abate them. In the case of cap-and-trade systems, they require strong regulatory

[16] RECs are used worldwide to promote renewable energy. In the USA and Canada, they help meet state or provincial regulations. In Europe, Guarantees of Origin (GOs) certify renewable energy, while Australia uses Small-scale and Large-scale Certificates (STCs/LGCs). Latin American countries have similar systems, with different levels of success (see details in Allub et al. 2024, Appendix to Chap. 4). Asia also has notable programs, especially in Japan, South Korea, and China.

frameworks, reliable emissions monitoring, and enforcement mechanisms. Emerging countries, in particular Argentina, face challenges with institutional capacity, transparency, and governance, making it difficult to establish and enforce an effective carbon market. Building the needed infrastructure is relatively costly and complex.

In the case of Argentina and several emerging economies, such as Mexico, Brazil, and Chile, a successful example was the relatively easy combination of elements that indirectly contribute to mitigation by promoting renewables. The combination of an RPS scheme with RECs and long-term auctions has helped support renewable projects by securing stable incomes for generators and giving market operators enough flexibility. Auctions for renewables were first implemented in Argentina in 2016 with marked success but ended in 2019.[17] During 2023, renewable generation tenders were reactivated through the nationwide RenMDI program. The initiative revolved around two objectives: on one hand, replacing "forced generation," i.e., inefficient power plants dispatched due to necessity rather than economic viability, including inefficient power plants located at isolated localities. In this regard, the tender promoted the replacement of these inefficient power plants by installing photovoltaic, wind, and biomass plants in regions with limited electricity supply. Secondly, RenMDI aimed to drive greater diversification of the energy matrix by leveraging the introduction of alternative technologies: biogas and landfill biogas plants, small hydroelectric plants, and biomass. The auction was successful in the sense that it managed to award approximately 100 generation projects to cover the 620 megawatts (MW) of capacity stipulated in the bidding terms of CAMMESA.

It is important to note that renewable auctions and RECs are not subsidy mechanisms and therefore do not create distortions. In that sense, the focus should be on using the "correct" market prices that incorporate all private costs plus the negative externalities of fossil fuel use and on having clear rules to ensure a favorable context for investments.

4.3.1.3 Improvements in the Industrial Organization of the Electricity Market

It is essential that electricity supply be governed by the economic dispatch order and have an independent system operator (ISO) to coordinate, control, and monitor market operations (Gutiérrez-Meave et al., 2021). Those are the roles for which CAMMESA was precisely created. Here we highlight two measures aimed at achieving greater efficiency in the electricity market, thereby reducing costs and

[17] Argentina's renewable energy auction policy, the RenovAr program, launched in 2016, initially attracted significant investment and competitive prices for wind, solar, and other projects. However, it was abandoned in 2019 due to several factors: a severe economic crisis with high inflation and rising debt, financing challenges for investors, a government shift toward stabilizing the economy, and currency exchange risks, as contracts awarded in U.S. dollars faced peso depreciation. These issues halted new auctions and delayed projects, with the abandonment driven more by macroeconomic instability than the program's effectiveness. In mid-2019, Argentina held the last RenovAr Round, the 3rd, with a minimum price of 54.22 USD/MWh.

improving service quality, which also contribute to the decarbonization of the economy.

Firstly, the incorporation of nodal prices could be used to provide a more accurate representation of the supply and demand dynamics at specific locations in the electricity grid. This allows for a precise reflection of transmission constraints and congestion. Nodal pricing also enhances market transparency by furnishing detailed information about prices at different nodes or locations, enabling market participants to make more informed decisions. Moreover, nodal prices encourage efficient resource allocation by reflecting the true costs of generation and transmission at different locations. This, in turn, leads to better investment decisions and optimal resource utilization. Market participants can manage risk more effectively through improved hedging strategies for generators, consumers, and other players in the market. Finally, nodal pricing signals can incentivize investments in transmission infrastructure in areas with congestion, promoting a more resilient and reliable grid.[18]

Secondly, focusing on the demand side, the sector needs a complete restructuring of pricing mechanisms for end consumers. Two forces are in effect. On the one side, this step involves transparently presenting electricity costs so that they are reflected in the rates for different user categories. Therefore, efforts should be directed towards ensuring that consumers react to price signals. The initial phase would entail installing smart meters that enable real-time pricing. While this implementation is not yet widespread in developed countries, it is undoubtedly the direction to pursue in the coming years. The goal is to align supply with demand, with the latter responding to real-time price signals (efficient social marginal costs). Naturally, this also necessitates providing users with information to comprehend price dynamics and adjust their consumption accordingly.[19] On the other hand, this need must balance with the real conditions under which generation capacity is promoted and remunerated. In the case of high Capex and low Opex that characterizes non-conventional renewable energy, competitive tenders are usually followed by average-price contracts, so that, even though marginal costs vary on an hourly basis—and they may fall to zero during long period of times—generators receive an average price paid by the contract counterparts, being distribution companies or large users.

---

[18] Implementing nodal pricing systems can be complex and requires sophisticated market structures, operational mechanisms, data requirements, and some extra administrative costs. Additionally, nodal pricing may create opportunities for market participants to exercise market power, especially in constrained areas, leading to higher prices and reduced competition. Nevertheless, Argentina is certainly able to effectively implement this more sophisticated system and has the tools and capabilities to avoid these unfavorable situations.

[19] Countries like Peru, Brazil, Costa Rica, and Uruguay have introduced intelligent meters, but Argentina is lagged behind. The case of Chile is interesting, for in the first attempt there was lack of acceptance by residential users because they did not want to pay for them, as the regulation stated. The second attempt was a self-selection mechanism between a Time of Use tariff and the standard tariff.

4.3.1.4 Support for Distributed Generation

Expanding on the discussion of clean renewable generation mentioned earlier, another crucial aspect is the support for the development of distributed generation. Specifically, solar photovoltaic energy presents a valuable opportunity to decentralize the processes of electrical energy production and consumption. Once again, specialized planning is essential, particularly for the distribution line, to ensure the reliability of the system and prevent setbacks.

Several compelling reasons support the idea that renewable energy should not exclusively focus on the development of utility-scale projects but also emphasize promoting distributed generation for small and medium users. Firstly, large-scale projects necessitate extensive land parcels, potentially valuable for other purposes, while distributed systems can be installed in various locations such as rooftops, parking lots, corporate yards, and more. Secondly, large solar farms currently face prolonged periods of navigating through numerous regulations and authorizations. Thirdly, environmental and wildlife protection groups have filed lawsuits against wind, solar thermal, and photovoltaic plants. Fourthly, distributed solar systems mitigate energy losses in generation, long-distance transmission, and distribution. Fifthly, distributed generation contributes auxiliary services to the grid, including voltage control, demand-side management to smooth utility system load peaks, and enhanced energy quality (e.g., batteries smoothing the variable output of a photovoltaic system). Sixthly, distributed systems have the potential to supply electricity during grid outages and resulting blackouts from unforeseen situations. Seventhly, decentralizing the network can potentially shield people in one region from issues affecting other areas, such as blackouts. Emerging technologies, including smart meters and small-scale battery storage systems, will further enhance this value. Eighthly, there is evidence that households are willing to pay for increased renewable and better-distributed energy (Martínez-Cruz and Núñez 2021; Hancevic et al. 2025). Lastly, while unrelated, distributed generation can significantly aid by providing reliable service in areas with challenging access to the public electricity grid, such as remote rural areas without access to the public grid.

Argentina's Law 27.424 on Distributed Generation, enacted in 2017, allows users to generate their own energy from renewable sources (such as solar or wind) and sell any surplus back to the grid. The law promotes the concept of "prosumers," who both consume and produce energy. Key provisions include net billing, allowing users to be compensated for energy fed into the grid, and various fiscal incentives to support the purchase of renewable energy equipment. Additionally, the law establishes a National Registry of User-Generators (ReNUG) to organize participation. Provinces such as Córdoba, Mendoza, Santa Fe, Buenos Aires City (CABA), San Juan, and Tucumán have adhered to and regulated the law, with varying levels of implementation. Other provinces are at different stages of compliance, with progress often depending on local regulations and agreements with electricity distributors.

Currently, the total cost of a fully assembled rooftop photovoltaic system in Argentina is approximately \$1.75 per watt of installed capacity.[20] This cost is double that in neighboring Brazil (\$0.8/Watt) or distant Australia (\$0.7/Watt). Nonetheless, costs in Argentina are nearly half of what they were just 15 years ago, complemented by the fact that panel efficiency is almost twice as high.

A positive aspect is that the mechanism to settle balance is based on net billing, overcoming the economic criticisms on former mechanisms based on net metering (see, e.g., López Soto et al. 2019). But one of the significant challenges lies in the absence of clear rules or uniformity in the implementation of distributed generation across different provinces of the country. Furthermore, there is a complete lack of awareness among users, resulting in an adoption rate that is nearly zero.

Ultimately, price emerges as the predominant factor. Since residential tariffs remain low or subsidized, the abundant availability of solar resources alone cannot ensure the viability of rooftop PV installations. Even in countries where regulations are sufficient and the interconnection process is relatively simple, such as Mexico, residential electricity rates are heavily subsidized in some districts, discouraging solar panel adoption (Hancevic 2017; Hancevic et al. 2022a, b).

4.3.2 Gradual Elimination of Fossil Fuel Use

The second pillar of the energy transition policy revolves around the substitution of fossil fuels, encompassing not only the previously discussed electricity generation but also the remaining economic sectors such as manufacturing, commerce, services, households, transportation, and the primary activities, i.e., agriculture, forestry, fishing, and mining. Electricity consumption represents only 20% of energy consumption, natural gas consumption by end users represents another 37%, while fuels represent 35%. A characteristic of Argentina is that coal is almost eliminated from the matrix. The elimination of fossil fuel usage promises several evident benefits, including enhanced public health due to cleaner air, water, and soil. Moreover, it offers advantages such as decreased reliance on fossil fuel imports, heightened energy sovereignty, and improved political stability (IRENA 2020a, b, d; Markandya and Wilkinson 2007). Achieving this ambitious objective necessitates structural reforms, which we will scrutinize in the following subsections.

Several policies exist in Argentina and other countries in the region that influence the perceptions and expectations of economic agents, leading to distorted energy prices (see, e.g., Hancevic et al. 2016). While energy subsidies may be justified from a distributive and anti-poverty perspective in some cases, their poor targeting—i.e., severe inclusion and exclusion errors—often results in a wasteful allocation of

[20] International Solar Energy Society (ISES) column in PV Magazine, March 24, 2024. Available at: https://www.pv-magazine.com/2024/03/22/pv-and-prices-the-not-so-fast-uptake-of-solar-in-argentina-and-chile/.

resources.[21] Furthermore, when the prices faced by the final consumer are distorted, there is an excess of consumption that impacts not only economic efficiency but also local pollutants and GHG (Di Bella et al. 2015; Parry et al., 2023; Parry et al. 2014). Ideally, energy prices should reflect all costs of supply, as well as the negative externalities caused by GHG emissions and local pollution.

4.3.2.1 Implementation of Market Instruments to Reflect the Price of Carbon and Local Pollutants

The electrification of extensive sectors, notably the industrial and transportation sectors, is not an instantaneous occurrence but rather a technological transition anticipated by most analysts. This transition is already underway, exemplified by the European Union's mandate to prohibit the manufacturing of internal combustion engine vehicles by 2035. Meanwhile, the most direct means to reflect the accurate cost of polluting fuels is through a carbon tax.

In the specific case of Argentina, the country introduced a carbon tax in January 2018 to address climate change and reduce GHG emissions. This tax targets the carbon content of fossil fuels, such as gasoline, diesel, fuel oil, and coal, following a "polluter pays" principle. Initially set at 10 USD per ton of CO_2 equivalent, it applied to 10% of the full value at launch, with gradual increases planned. Natural gas used for electricity generation and biofuels were exempted to avoid impacting energy prices and promote renewables. Revenue from the tax contributes to the national budget and can be directed towards environmental programs. While the carbon tax aligns with Argentina's commitments under the Paris Agreement, its low rate may not significantly deter fossil fuel use. Economic challenges, including inflation and recession, further complicate its implementation and public acceptance. Nevertheless, the tax represents a step towards a low-carbon economy, aiming to stimulate investment in renewable energy.

It is worth emphasizing that all applicable taxes (i.e., existing levies like value-added tax, sales tax, or gross income tax) should be considered as a complex mechanism of carbon taxes (see, e.g., Belfiori and Rezai 2024). These taxes abound at both national and local government levels in Argentina. The crux is that the production price plus the sum of levies should mirror the social marginal cost of the fuel, amalgamating all private supply costs and the social cost of the negative externality stemming from its production and consumption. The ensuing rise in consumer prices will invariably expedite the energy transition, compelling Argentina to actively engage in this transformative process.

In the theoretical realm, the implementation of a carbon tax is equivalent to establishing a comprehensive emissions permit market—cap-and-trade—encompassing

[21] Hancevic et al. (2016) describe cross-subsidization as prevalent (and often widespread), with price cycles separating from long-run sustainable marginal costs. In addition to being a fiscal problem, these situations threaten the reduction of GHG emissions, the adoption of distributed generation, and investments in energy efficiency improvements and equipment replacements.

all industries, including electricity generation. In general, market mechanisms (both taxes and cap-and-trade systems) have proven more efficient than command-and-control policies in curbing emissions (Hancevic 2017; Blackman et al. 2018).[22] The underlying advantage of market-based environmental regulations is their ability to achieve a specific environmental objective at a lower cost, especially when agents are heterogeneous in terms of their marginal abatement costs. Also, market-based mechanisms promote GHG reduction through intermediate effects from both technological progress and energy consumption structure perspectives. In contrast, command-and-control regulations primarily achieve this through the former avenue (Guo et al. 2021).

For the specific case of Argentina (and perhaps for most emerging countries), the institutional landscape favors the imposition of generalized taxes (for carbon and local pollutant) over a cap-and-trade system. Taxes, coupled with sector policies such as energy efficiency or incentives to introduce a certain technology, are more easily navigated through legislative bodies and are more defensible against public opinion. This proposition aligns with the tariff rebalancing and cost transparency of the electricity market discussed earlier. In both instances, the crux is to reflect the social marginal cost of energy.

4.3.2.2 Substitution of Polluting with Non-Polluting Transportation

The benefits of substituting polluting private transportation with non-polluting alternatives are self-evident. Reduced reliance on private motor vehicles leads to diminished health system costs owing to decreased pollution and fewer traffic accidents. Additionally, it mitigates congestion and reduces travel time (Calatayud 2021). The challenge lies in fostering suitable infrastructure that promotes "active mobility" such as bicycle ways, pedestrian walkways for safe commuting, and efficient public transportation lines. Conversely, concerted efforts should be made to discourage the usage of private motor vehicles. Acknowledging that the transportation sector is the primary emitter of GHG in LAC among energy consumers and anticipating an increasing reliance on clean renewable energy sources in electrical generation, the transition to a fleet of electric vehicles and an electricity-powered public transportation system represents a noteworthy advancement in decarbonization. Bogota (Colombia) and Santiago (Chile) are leading cases in the introduction of electric buses. Buenos Aires is just in the preliminary phase of bus replacement. The associated benefits align with those discussed earlier.

In many Latin American and Caribbean (LAC) countries, the railway and river transport infrastructure remain underdeveloped or underutilized, with road trucks predominantly used for cargo transportation. Argentina serves as a paradigmatic case, having gradually abandoned its once cutting-edge railway network in favor of

[22] Command-and-control measures include emission standards, fuel consumption requirements with specific characteristics (e.g., certain levels of sulfur content), and the obligation to install decontamination technologies, among others.

fleets of diesel trucks. Transitioning to electric trains for heavy freight transport offers a viable alternative. Currently, rail freight represents about 5% of the modal share, through proper incentives, could increase up to 10–15% in the medium run. Similarly, since the country has plenty of navigable rivers and lakes, it should certainly prioritize transportation via boats and barges. Additionally, for shorter distances, land transport through electric battery trucks could be a medium- to long-term solution. Finally, the integration of green hydrogen as a tangible alternative for powering passenger and freight transport engines remains an aspect to be explored, given the enormous potential the country possesses.

4.3.2.3 Industry Electrification and the Use of Green Hydrogen

When referring to the industrial sector, we encompass various sectors emitting GHG. These range from steel, cement, and chemical industries to generic processes that generate heat at highly variable temperatures, depending on the specific needs of each industry. The benefits of electrification can be substantial for society if most of the electricity comes from clean, renewable energy sources, which, as mentioned earlier, are now cheaper to produce.

For the more specific case of heavy industries, which usually require higher temperatures in their processes (e.g., cement, metallurgy, chemical, and fertilizer industries, etc.), electrification could be difficult or very costly to implement. But there may be excellent opportunities for the development of gas as a transition source, and green hydrogen as a substitute for coal and fossil fuels in the medium term, providing clean high-temperature heat.

In summary, regarding the elimination of fossil fuel use, the debate at the governmental and societal level should address the following. First, how can the path be smoothed by creating clear institutions and rules so that, through the efficient operation of markets—i.e., with the "correct" social prices of energy, without information barriers, anticompetitive practices, or other market failures—, the desired energy transition can take place? Second, not only does the final result matter (achieving zero emissions by 2050), but the path and the timing of this transformation also do. For example, it is expected that vehicle terminals will stop producing units with engines using fossil and polluting fuels in the coming years. So, how timely would it be to facilitate the transition to electric vehicles so that Argentina becomes a key player in this industry and does not lag behind other regions of the world? There is plenty of history of automobile terminals in this country that could be applied to electric cars. It remains to be determined what impetus the government needs to provide with public funds to support the supply side in these industries, without displacing private investment with public investment. Second, are there market failures on the supply side that, if corrected, could lower costs, for example, through learning-by-doing and positive externalities among sector producers? Similar dilemmas occur on the demand side. Following the example of electric vehicles, is it necessary for the state to promote improvements in EV charging infrastructure, or is it in the best interest of automobile manufacturers to invest in such network-like infrastructure?

Again, the next question is: to what extent should private development take the lead, or is it possible to accelerate individual adoption, thereby providing more incentives for manufacturers to invest in infrastructure?

4.3.3 Complementary Measures

4.3.3.1 Energy Efficiency Programs

In terms of energy efficiency, the country lacks a comprehensive national regulation, despite partial laws and initiatives covering labeling in homes, equipment, SMEs, and lighting. The National Program for Rational and Efficient Use of Energy, established by Decree 140/2007, stands out as a significant effort. The Energy Efficiency Labeling Program (2007) introduced the labeling system and minimum standards, mandating labels for various appliances. However, labeling for other equipment remains voluntary. Law 26.422 in 2008 provided the legal framework for the Secretary of Energy to set standards for machinery. In 2021, Resolution 383/2021 from the Ministry of Environment and Sustainable Development progressed in CO_2 and vehicle energy efficiency labeling.

Regarding private residences, the National Housing Labeling Program (PRONEV) aims to implement a unified system nationwide, but implementation is gradual. The Efficient Lighting Plan (PLAE), initiated by Resolution 84 of 2017 by the former Ministry of Energy and Mining, encourages LED technology in public spaces. Provinces and municipalities receive non-refundable contributions for LED replacements, subject to specified criteria. A 2021 technical update involved joint consultation with INTI and others to align specifications with technological advancements. This plan aligns with laws 26.473 and 27.492, prohibiting incandescent and halogen lamps while promoting LED technology.

What we commonly observe are isolated and entirely uncoordinated efforts that reflect good intentions but do not ensure effective results. Energy efficiency programs, when properly implemented, have the potential to reduce energy consumption. However, achieving this goal requires overcoming informational barriers and addressing other market failures and constraints, including insufficient financing available to low-income users, price signals and subsidies that are currently misaligned with energy efficiency. Although these energy efficiency measures alone may not lead the economy directly to the desired energy transition, they can significantly contribute to enhancing the well-being of the population by improving allocative efficiency. Thus, these initiatives should be seen as complementary measures to those discussed throughout the chapter.

4.3.3.2 Demand Management Strategies

Demand response programs, representing the actions that users of the electrical system can undertake for the benefit of the supply system, constitute a critical component of energy economics. Historically, the primary action considered within these programs is self-reduction of demand to mitigate the risk of major system collapses. This proactive measure yields economic benefits for the user. In Argentina, such actions have been integrated into the standard operating procedures of the system for several decades, encompassing mechanisms such as Interruptibility, Instantaneous Reserve Service, and participation in the Load Relief System. However, these initiatives often fail to appeal to typical users due to remuneration values being either negligible or non-compatible with their interests or simply due to a lack of awareness regarding the existence of these practices.

Therefore, as an initial course of action, regulatory authorities should update the remuneration rates for these services, aligning them with competitive standards so that they reflect scarcity conditions. Furthermore, advancing the development of these services entails the progression of storage systems, including end-user facilities. Presently, such advancements facilitate demand management to extend its role beyond mere demand reduction interventions, allowing for proactive energy contribution to the grid as per its exigencies.

This capability would enable the agent (whether it be a transporter, a distributor, or an end-user) to engage in other services such as primary and secondary frequency regulation, peak load shaving, forced generation, and essentially the equivalent services of a generator. These services will need remuneration commensurate with the level of investment required, considering that not only must a storage system be available, but also communication systems conducive to operation, power electronic equipment facilitating such operation, and among others.

4.3.3.3 Network Infrastructure and Digitalization

The electric grid stands as pivotal infrastructure in all societies, gaining increased importance during the energy transition. A modern and intelligent grid, integrating information technologies and the Internet of Things, promises enhanced returns on investment. Smart meters, central to smart grids, facilitate bidirectional energy flow measurement, enabling distributed energy, efficient demand management, and electromobility services through grid-connected bidirectional chargers. Upgrading the grid for intelligence and resilience, employing advanced technologies, equipment, and coordinated controls, ensures a more reliable and efficient electricity supply. This translates to improved user services, including notably reduced power outage frequency and duration, and expedited service restoration during interruptions.

Energy consumption varies over time and is not uniform. The distribution network must expand to accommodate the maximum expected demand in the future. Traditionally, this design has been based on past consumption experience. However, substantial changes in user consumption profiles are expected in the future, making it crucial to anticipate these changes to optimize investments. The implementation of smart meters aims to provide this information. Concretely, a massive replacement of existing traditional meters with smart meters is needed, which could be financed, for instance, through a specific charge on electricity demand. Although the electricity grid is updated with technological advances (centralized control is highly advanced), Argentina is still at early stages of innovative automated measurement infrastructure.

4.3.3.4 Biofuels Development

In Argentina and globally, biodiesel and bioethanol represent the most developed forms of bioenergy, primarily used in transportation where they are commonly blended with other fossil fuels. Nevertheless, there are other types of secondary biomass being harnessed for energy purposes, albeit on a smaller scale and with less advanced technological developments compared to the aforementioned energy sources. On the international stage, the biofuels market is usually subject to stringent regulations. The mandatory blending of gasoline and diesel with biofuels serves as a catalyst for their production or importation. In major industrialized countries, robust state subsidies and tariff barriers exist to incentivize and bolster biofuel production. The Argentine scenario mirrors this pattern, as its expansion is closely tied to the active involvement of the State.

Enacted in 2006 and updated in 2021 by Law 27,640, Biofuels Law 26,093 establishes a Regulation and Promotion Regime for Sustainable Biofuel Production and Use until 2030. As of October 2023, biofuel incorporation is set at a minimum of 7.5% biodiesel in diesel (temporarily 12.5%) and 12% bioethanol in gasoline. For sugarcane- and corn-based bioethanol, volumes must be 12% of the mandatory minimum blend, according to Resolution 631/2023 from the Ministry of Energy.

What must be clear is that any strategy supporting biofuels should be viewed as short- and medium-term measures, releasing pressure on land use and food security. The long-term envisions the complete removal of fossil fuels, either through significant progress in electrification or the emergence of green hydrogen, indicating that biofuels will likely play a role in the transition.

4.3.3.5 Natural Gas: The Transition Fuel

The increasing demand for energy worldwide suggests that the transition will be neither simple nor linear. The primary objective is to reduce coal consumption, a reality recognized by various international taxonomies as they incorporate natural gas as a transitional fuel. The next step is to reduce oil-based fuel consumption. In this context, natural gas can be considered the interim fuel during the Argentinean energy transition for several reasons. First, compared to traditional fossil fuels like coal and oil derivatives, natural gas generally produces lower carbon emissions when burned for energy (Sepulveda et al., 2018). This makes it a cleaner option, helping to reduce the overall carbon footprint during the transition phase. Second, Argentina already has a well-established infrastructure for natural gas extraction, transportation, and distribution. Utilizing this existing infrastructure can provide a smoother transition compared to introducing entirely new technologies or energy sources. Third, natural gas is versatile and can be used for various purposes, including electricity generation, heating, and as a fuel for vehicles. This flexibility allows it to meet multiple energy needs during the transitional period. Fourth, natural gas can complement intermittent renewable energy sources like wind and solar power, at least until other cleaner technologies become competitive. Additionally, it can serve as a reliable backup during periods of low renewable energy production, ensuring a more stable and consistent energy supply. Fifth, transitioning directly from traditional fossil fuels to entirely renewable sources may pose economic challenges. Natural gas provides a more economically viable and readily available option for meeting energy demands while gradually incorporating renewable technologies.

The Vaca Muerta field represents a unique resource location for Argentina, offering a window of development opportunity that contributes to the decarbonization of regional and global economies. The natural gas resources and reserves of Vaca Muerta, coupled with the productivity and estimated ultimate recovery of its wells, ensure a stable supply. Thus, if the country implements the necessary measures to create market conditions conducive to resource development, production could potentially double, reaching high levels of exports and generating high-quality employment.

However, it is crucial to note that the role of natural gas in the transition should be seen as an intermediate step towards a more sustainable and renewable energy future. The ultimate goal is likely to transition towards cleaner and renewable energy sources as technological advances and becomes more economically viable.

4.3.3.6 Just Transition

The Paris Agreement recognizes that the efforts to reduce temperature through greenhouse gas emissions should be done in a context of sustainable development and recognizing national circumstances. The idea of just energy transition is a recognition of several heterogeneities at an intergenerational level (not leaving future generations unaccounted for), among countries (i.e., reduce the economic gap with the developed world) and citizens (i.e., reduce social and energy inequalities), as emphasized by Cont et al. (2022) and Allub et al. (forthcoming).

The literature has studied the relationship between the economic development path and emissions, giving rise to the hypothesis of an inverted U-shaped relationship, which is called the environmental Kuznets curve (see review by Kaika and Zervas 2013). The argument for Latin America, and for Argentina in particular, is that most countries did not reach a level of development to engage on a path of decreasing emissions with growth. In 2019, per capita GDP in Argentina (in constant values) was about 64% of per capita GDP of the OECD group, while primary energy supply was 77% of OECD's primary energy supply. The values for Latin America were 42% and 51%, respectively.

Argentina, as many other Latin American and Caribbean countries, did not overcome the development threshold yet. In the second half of 2023, the poverty rate exceeded 40%. This dimension of development is still unresolved in our country, and in all the region, and is a source of additional condition for energy transition policies.

A third source of heterogeneity is related to the location of natural resources. While countries such as Uruguay or Chile do not have natural energy resources in its primary matrix and clean generation capacity grew fast, other countries such as Argentina, Colombia, Ecuador, Mexico, and Trinidad and Tobago, depend at different levels on resources, not only as economic activities but also as sources to the fiscal accounts. An abrupt change in the supply matrix led by the energy transition may affect them. Of course, there may be significant compensations if the transition moves from fuels to lithium in the country, but this puts another constraint into play. Specifically, the Vaca Muerta field is in Neuquén province while lithium resources are located in Jujuy and Salta. The reliance of provinces on natural resources-based royalties is an important consideration to take.

On the other hand, a just transition should encompass policies that support workers and families whose jobs, incomes, and livelihoods are at risk as the world seeks sustainable paths (Rosemberg, 2017; Graham and Knittel 2024). A policy that encourages the retraining of individuals for the transition to renewable energy industries would also provide a just transition for workers currently employed in fossil fuel-based industries (Saget et al., 2020). A shift to the informal sector is undesirable because it would decrease wages in the informal sector—negatively impacting not only newcomers but also those with permanent informal employment, and it implies a loss in the returns on individuals' and society's investments in skills that are not useful in the informal sector—a waste that is difficult to recover from once workers transition to the informal sector because they may not easily return to the formal sector (Levy 2020). The benefits of electrification with a just transition may be more substantial for society in the future than merely considering it from an energy perspective (Martínez-Cruz and Núñez 2021).

4.3.3.7 Resilience to the Impacts of Climate Change Through Green Growth Strategies

Examples of these policies that can contribute to promoting clean energy and enhancing climate resilience encompass a variety of financial instruments. Targeted loans, for instance, involve directing financial resources to specific projects or initiatives related to clean energy, allowing for a more focused and effective impact. Green bonds, another notable example, involve issuing bonds specifically designated for financing environmentally sustainable projects, such as renewable energy infrastructure or energy efficiency initiatives, usually tied to reduced-rate incentives to meet green targets (IRENA 2020c). Moreover, fiscal credit policies designed for investors or consumers play a crucial role in incentivizing the production and utilization of clean energy. These policies may include tax credits, subsidies, or other financial incentives that encourage individuals and businesses to invest in, adopt, or transition to cleaner and more sustainable energy practices. By providing financial benefits, these measures help offset the costs associated with adopting cleaner technologies, making them more attractive and economically viable. Lastly, the establishment of regional and national climate funds serves as a pivotal strategy to mobilize, access, and channel financing for climate-related projects. These funds are dedicated to supporting initiatives that contribute to climate change mitigation and adaptation efforts. By centralizing resources at both regional and national levels, these climate funds facilitate the efficient allocation of financial support to projects that address environmental challenges and foster sustainable development.

However, it may take time until Argentina is accepted back in the international market. Meanwhile, multilateral development banks may play an important role both as financers of green projects and as catalysts of available climate funds.

4.4 Challenges

The key challenges that impede Argentina's progress towards a sustainable energy transition are the distributive impact, the outdated infrastructure, the pervasive influence of energy subsidies, and the financial constraints and regulatory hurdles.

4.4.1 Distributive Impact

The line of reforms recommended in Sect. 4.3 has a high component of economic incentives aimed at improving energy efficiency, reducing market frictions, and fundamentally reducing GHG emissions. The economic and environmental benefits are evident. The effort of the private sector is considered through a "facilitator" state that provides efficient market conditions, ultimately encouraging the adoption of clean and sustainable energy. However, the restructuring of markets and the accounting of "correct" energy prices may have certain negative effects on the well-being of some consumers. This is not something new, and, in fact, it is common to almost all reforms and political measures that governments can implement. In the case of the energy sector, it could happen that the least privileged sectors suffer negative changes in well-being. Unfortunately, households below the poverty line represent a considerable part of society in Argentina (currently about 40% of the population), and there is great income inequality. However, most studies that have analyzed how the adjustment of energy prices through the removal of subsidies and the application of carbon taxes could affect households have reached two fundamental conclusions. First, the final impact of the reform depends on how the additional tax revenue generated is used. Second, only a portion of the additional income from the reform is needed to compensate for the negative changes in the well-being of vulnerable households.[23] For example, the undesirable situation could be corrected in electricity consumption through tariff structures tied to correctly targeted social tariff schemes. Thus, initially, prices should be brought to sustainable long-term levels, that is, levels compatible with social marginal costs that include not only comprehensive production costs but also the costs of pollution externalities. As a complementary measure, and assuming that the country's laws allow it, subsidy mechanisms could be implemented so that less favored sectors receive fixed-sum support that does not distort the decisions of agents.[24]

The correction of electricity tariffs and fossil fuel prices is a step forward towards economic efficiency but also towards energy transition. It is expected that in this context, better decisions will be made regarding the replacement of obsolete equipment, investments will be made in energy efficiency improvements, clean distributed generation systems and solar water heaters will be adopted to a greater extent, internal combustion vehicles will be replaced by electric vehicles, industrial electrification will be achieved, among others. In countries where current tariffs are heavily subsidized, the savings from subsidy correction could even be redirected towards implementing subsidies to finance the adoption of solar panels or investments in energy efficiency. In other words, inefficient current energy consumption subsidies can be temporarily replaced by capital subsidies to finance the adoption of green technologies. Empirical evidence tells us that currently, the adoption of green technologies is biased towards middle-high and high-income households (Ameli and Brandt 2015; Gao and Zhou 2022). Therefore, capital subsidies focused on more vulnerable households could be a superior alternative. All of this should be accompanied by interventions that affect user behavior, eliminate information problems, and other market failures to achieve sustainable social benefits over time.

On the other hand, the gradual disappearance of the fuels and hydrocarbons sector in the long run could leave stranded assets and lead to people connected to those industries losing their jobs, creating friction in the labor market when relocating those workers. It is necessary in that case for governments to proactively (not reactively) assist in facilitating the transition and displacement of workers between industries. At first, job generation linked to the booming renewable energy sector is very significant. Globally, it is expected that the number of jobs in the energy industry will grow from around 18 million currently to more than 26 million in 2050, with 84% of them linked to renewable energies, 11% to fossil fuels, and 5% working in the nuclear energy subsector (Pai et al. 2021).

[23] See, for example, Symons et al. (2002), Bento et al. (2009); Burtraw et al. (2009), Parry and Williams (2010), Liang and Wei (2012), Feng et al. (2018).

[24] Hancevic et al. (2022a) propose a means-tested mechanism by simulating a census of the poor, enabling efficient tariff rebalancing for residential electricity users in Mexico. Cont and Navajas (2019) emphasize the need of tracking reforms, so beneficiaries do not disconnect the subsidy from the tariff adjustment.

4.4.2 Infrastructure

The infrastructure needs are numerous and in various sectors, costly, and unavoidable. In the case of the electrical grid, investments are required both in clean generation facilities and transmission and distribution lines. With regard to generation, the private sector has the main advantages (capital and knowledge) and therefore clear rules of the game are required. Market mechanisms such as the long-term auctions are likely necessary for the development of this sector. In addition, regional integration offers an opportunity to diminish the requirement for additional generation capacity solely for peak demand coverage. Through collaborative efforts with neighboring nations, a surplus of firm capacity can be shared, reducing the burden on individual grids. Advancing towards deeper integration could unlock significant benefits. By leveraging geographical diversity, it is projected that at least 2.5 GW of intermittent renewable power could be integrated without the need for backup support. Additionally, this integration could curtail the necessity for thermal backup.

In the case of the transmission network, it will need to expand significantly to integrate the increased demand resulting from the massive electrification process. Furthermore, the inclusion of a sizable quantity of solar and wind energy capacity implies even greater investments in transmission infrastructure and reliable backup. As electricity demand rises, there is a pressing need to expand transmission capabilities across the national landscape, particularly from regions rich in renewable resources. This strategic investment not only addresses the imperative for enhanced infrastructure but also alleviates the reliance on gas pipelines outlined in the interim (transition) period. Secretary of Energy (2022) has an updated map of short- and medium-term bottlenecks. In the long term, to meet the electricity transportation needs in a green scenario as outlined here, it is estimated that by 2030 it will require an investment of $16 billion, and by 2050 around $46 billion. These figures imply an addition of 32,000 km of additional transmission lines (Deloitte–Enel 2023).

A fraction of these investment needs could be delayed if the interconnection infrastructure is optimized for regional use. However, the contract breaching events that took place at the advent of the twenty-first century require that Argentina renews commitments to improve relationships with partner countries.

Continuing with infrastructure needs, while it is important to involve private initiative, centralized planning through a system operator that bases investment needs on, for example, a nodal pricing mechanism, is crucial. It involves anticipating network congestion situations as more sources of clean generation are incorporated and as electrical demand expands.[25]

[25] The case of Chile is interesting in the sense that, even with significant investments in transmission, the result net is a reduction in the final cost of electricity for users. In recent years, major works have been carried out to expand the transmission network, enabling the interconnection of the Northern Large Grid System (SING) with the Central Interconnected System (SIC). This market integration has facilitated exponential growth in renewable energy in the Atacama Desert (northern Chile), where clean energy is generated to supply the central and most populated area of the country (Gonzáles et al. 2022).

The substitution of fossil fuel use in the field of transportation requires certain initial investments. On the one hand, it is necessary to expand coverage and improve the quality of urban public transportation to make it a genuine alternative for users. Often, the service is not widely used due to safety and quality issues. The barriers are probably behavioral: potential users opt for other forms of private transportation over public transportation. All this lies within the local (urban) planning realm, where various common objectives can intertwine, such as urban planning, congestion and travel time savings, safety, health, etc.

Barriers to electric vehicle (EV) adoption include the current price advantage of internal combustion engine (ICE) vehicles, though projections suggest EVs may become more cost-effective by the mid or late this decade due to expected battery price reductions (Cole et al., 2021). Despite the estimated lower maintenance costs and various benefits of EVs, such as improved range and quieter operation, consumer awareness remains low. Additionally, a lack of fast-charging infrastructure in public places hinders EV adoption, particularly in emerging countries. For example, there are scattered charging places in Ciudad de Buenos Aires and Greater Buenos Aires, but there is not yet a developed network connecting this region with main or popular cities, such as, Córdoba or Mar del Plata. The small EV market results in limited production by car manufacturers and higher import costs. These factors, coupled with fiscal concerns related to fossil fuel taxes and implicit subsidies, impede sector development. Government intervention is crucial in overcoming these barriers (Rapson and Muehlegger 2021; Quiros-Tortos et al. 2019). Measures may include facilitating charging infrastructure, promoting EV manufacturing, creating exclusive lanes, adjusting electricity rates, and providing financial incentives. However, a key consideration in these discussions is the identification of market failures justifying government intervention. Questions arise about subsidies for EV production based on learning-by-doing or positive externalities, the significance of network effects in public charging infrastructure, and whether EV manufacturers should bear these investment responsibilities. Addressing information gaps and correcting fossil fuel prices to reflect social costs are pivotal for consumer adoption decisions.

Many of these barriers also apply to freight transportation, specifically, those related to the charging infrastructure for electric vehicles (EVs). On the other hand, investment in improvements to railway infrastructure requires a shift in government strategy, turning a historical trend of (at least) the last 50 years. Currently, the budget is predominantly allocated to road construction. Electrification of port and railway infrastructure is necessary.

Many industrial processes demand substantial levels of energy and heat, posing challenges for electrification. Natural gas could be an interim source of energy, but not completely feasible in hard-to-convert sectors. Green hydrogen (H2) emerges as a clean energy carrier, offering a solution to address the issue of energy intensity, particularly for processes requiring high-temperature heat. Green H2 can be generated using renewable electricity through electrolysis. The intermittent nature of renewable energy sources can lead to inadequate and unreliable power for continuous industrial processes. In this context, green hydrogen functions as an energy storage solution. Surplus electricity generated during periods of high renewable energy production can be utilized for hydrogen production through electrolysis. The produced hydrogen can then be stored for later use during periods of low renewable energy generation. As the cost of renewable energy continues to decrease, green hydrogen becomes an increasingly competitive option for the decarbonization of industrial processes. Some industrial processes face obstacles due to the absence of mature electrified technologies, impeding the transition to electrification. Green hydrogen steps in as an energy carrier in sectors where direct electrification proves challenging. It can be employed in fuel cells or serve as a feedstock for industrial processes that are difficult to electrify directly. Retrofitting existing industrial infrastructure for electrification often necessitates substantial investments and presents technical challenges. Green hydrogen offers smoother integration into current industrial processes compared to direct electrification. This flexibility allows for adoption without the need for extensive modifications to existing infrastructure.

However, progress has been scanty. There has been a pilot project by the company Hychico (2008) in the province of Chubut, with two electrolyzers and a total capacity of 120 Nm^3/h of high purity H2, which is mixed with natural gas to power a 1.4 MW generator. Also, the R&D company related to YPF (Y-TEC) launched the consortium H2ar in July 2020, to promote the development of hydrogen production in the country. The announcement by the Australian company Fortescue, within the framework of COP 26, to develop a green H2 plant in the province of Río Negro to feed a 650 MW (expandable to 8GW) power plant, including a wind farm and a transport port, was not concreted yet, pending a regulatory framework for the sector.

4.4.3 Financial Constraints and Regulatory Hurdles

In this subsection, we assess two specific constraints. On the one hand, the transition involves large amounts of money. The public sector in the region, and specifically in Argentina, is far from being in condition to finance the transition. Argentina's public debt is over 80% of GDP and expected to rise. Fiscal accounts have run on deficit since 2010 (6% of GDP in 2023), although the balance is expected to revert for the first time in 2024. Therefore, the financing of investment should rely on the private sector, multilateral assistance, and alternative sources of funds.

On the other hand, general and sector regulations should be adapted to the energy transition. In the general case, it will help to give proper assurances to foreign direct investment (FDI) and local capital. In the specific case, the electricity regulation should adapt to the new setup (smart grids). Importantly, the electricity regulation is considered one of the more advanced in the region (World Bank, 2017), once provisions for intervention, introduced back in 2003, are removed. The index of sectoral reform of 2015 puts Argentina among the advanced countries (index of 96), well over Latin America (index of 72) and the world (index of 38). Of course, the legal framework needs adaptation and modernization (see Sect. 4.5).

In the case of the natural gas market, the countries that have the most developed markets at the end consumer level are Argentina, Brazil, Colombia, Mexico, and

Peru. In LAC, there are also countries with large production (Bolivia and Venezuela) or that are basically importers (Chile, Uruguay, and some Caribbean countries) but do not have a high development of the local market.

4.4.4 Energy Subsidies

Another important institutional aspect of the energy sector is the tariff and subsidy policy. In the context of the energy transition, one of the main challenges from a tariff perspective is related to the price level, accompanied by the policy of sectoral subsidies, since in many countries these do not cover the economic costs of providing the service (Cont et al. 2021). Cost coverage under efficient provision is essential for the economic and financial sustainability of the sector and allows the identification of constraints that countries may face when updating their energy policies. For example, undertaking a transition from petroleum products to electricity may create friction in a country if the consumption of the former (petroleum products) is subsidized. When the electricity or natural gas service is also subsidized, another dimension is added, as a subsidy in the sector implies consumption above the efficient level, which can make demand-side management policies (such as energy efficiency) more difficult to implement.

Argentina has a long history of energy subsidies. According to the International Monetary Fund (IMF 2023), direct subsidies through state contributions represented 0.7% of the region's GDP in 2023, compared to 0.8% worldwide (see Fig. 4.8). However, in Argentina, this figure increases to 1.6%, with oil derivatives and coal representing 9% (0.14% of GDP), natural gas 67% (1.07% of GDP), and electricity the remaining 24% (0.39% of GDP), confirming that subsidies are present at both ends of the transition.

Direct subsidies arise from the difference between the price charged by suppliers and the price paid by consumers. However, the supplier price does not necessarily reflect the total costs, including externalities such as environmental impacts (e.g., GHG emissions, local pollution, congestion). The IMF also estimates implicit costs to calculate total subsidies. In 2023, implicit costs represent 4.6% of LAC's GDP, compared to 5.7% worldwide (Fig. 4.8). In Argentina, implicit subsidies, which are more biased towards oil, coal, and natural gas, are aligned with the global level.

4.5 Opportunities[26]

Argentina possesses a plethora of renewable resources, including solar, wind, hydro, and biomass. These resources present a significant opportunity for Argentina to transition towards a more sustainable energy future. By harnessing these abundant

[26] This section is an update of Chap. 2 in Cont et al. (2022).

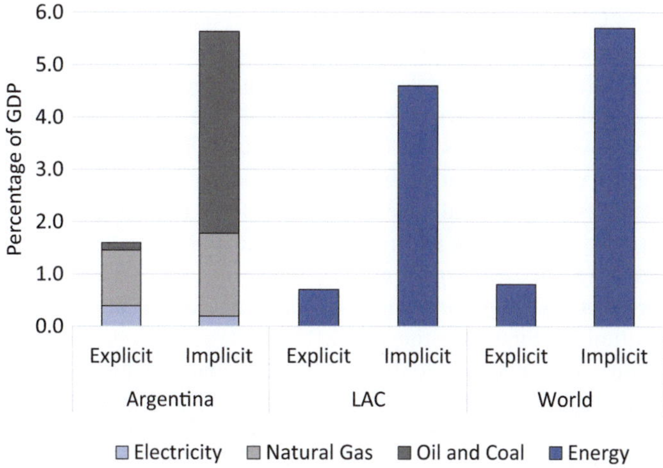

Fig. 4.8 Direct and implicit energy subsidies in Argentina, LAC, and world, as % of GDP of 2023. *Source* Own elaboration based on data from IMF (2023)

resources, Argentina can reduce its reliance on fossil fuels, mitigate GHG emissions, and enhance energy security. Moreover, technological advancements in renewable energy technologies, such as improvements in solar panel efficiency and wind turbine design, have made renewable energy more affordable and accessible than ever before. By leveraging these technological advancements, Argentina can develop its renewable energy infrastructure more efficiently and cost-effectively. Additionally, the prospect of global collaborations presents another avenue for Argentina to accelerate its energy transition. Collaborating with international partners, such as neighboring countries, international organizations, and multinational corporations, can provide access to expertise, financing, and technology transfer opportunities.

In detail, there are significant opportunities in:

- **Increase in the share of NCREs**. The penetration of non-conventional renewable sources (solar and wind) brings business opportunities and challenges of how to resolve their intermittency in electricity generation cycles. Greater connectivity, storage, and demand response will be required, which will necessitate increased investments in backup capacity and grid infrastructure. Moreover, the installation of new generation sources located far away from the existing grid will require extensions of transmission networks (or reinforcement if they already exist).
- **Development of hydrogen**. Hydrogen can involve major changes to the sector's production processes, including low-emission hydrogen production processes, transport, and storage systems. Currently, the production of green hydrogen is generally not competitive but is progressing through the learning curve.
- **Natural gas as a transition resource**. Beyond NCREs and hydrogen, natural gas may be useful in this substitution process because, in some cases, it produces up

to 50% less GHG emissions than other hydrocarbons (as long as methane emissions are contained), in addition to contributing to lower local pollution. Moreover, the availability of natural gas reserves in Argentina highlights its importance in ensuring energy security, especially in contexts with high penetration levels of renewable energies and the issue of intermittency. While NG may play a transitional role, stringent global decarbonization goals are likely to significantly reduce demand and revenues in the long run (Welsby et al., 2021). Finally, strong reliance on hydrocarbons could pose fiscal risks as the world decarbonizes (Solano-Rodríguez et al., 2021).

- **Electrification of energy consumption**. Transport electrification faces three business opportunities. First, investing in R&D for batteries; second, the deployment of infrastructure for connection to the grid for charging and discharging; and third, the interoperability of charging electric vehicles. There are also electrification alternatives for certain industrial, commercial, and residential uses.
- **Energy efficiency**. Improving energy efficiency requires three major changes. First, the infrastructure needs to be improved to make the transformation and distribution processes more efficient, with fewer losses. Second, it is important to assess instruments enabling the alignment of incentives to foster efficient use of energy by consumers. Third, investment in research and development must be encouraged to continue to improve the transformation and consumption processes through new technologies.
- **Changes in availability of sources due to climate change**. The greater frequency of extreme events, particularly droughts, may affect hydroelectric generation capacity, suggesting the need to diversify the energy matrix. Relying heavily on this source puts the system at risk in the face of projected water shortages.[27] Likewise, biomass production and availability may be affected by the adverse impact of climate change on agricultural activity.
- **Regional integration**. In response to issues like renewable generation intermittency and resource scarcity, regional integration emerges as a potentially viable solution for increasing countries' energy security. However, it poses major challenges regarding coordination and cooperation among the different States (a topic analyzed in detail in Chapter 5 of Sanguinetti et al. 2021).
- **Distributed generation**. Increased availability of renewable sources also facilitates the expansion of distributed generation. This paradigm shift is driven by improved technology efficiency and reduced costs, primarily in solar photovoltaics, which constitute nearly 98% of distributed generation installations in the region due to their characteristics and urban integration capabilities. Implementing such systems entails multiple considerations, including compensation schemes (energy or cash-based), minimum technical requirements to ensure the

[27] This argument may be more relevant in countries like Brazil or Colombia, where hydroelectric power dominates energy generation. However, in Argentina, hydroelectricity accounts for 15% of current energy production, and building new dams or upgrading existing ones could become an important consideration.

quality of distributed generation, the rate at which credits are exchanged with the grid, financial mechanisms (if any), and how to finance them.

By embracing these opportunities, Argentina can not only enhance its energy security but also position itself as a key player in the global sustainable energy landscape. Becoming a leader in renewable energy production and innovation can bolster Argentina's international reputation, attract investment, and stimulate economic growth. Furthermore, by demonstrating a commitment to sustainability and environmental stewardship, Argentina can contribute to global efforts to fight climate change and create a more sustainable future for generations to come.

4.6 Conclusion

The formulation of energy policies necessitates a comprehensive plan that facilitates the exploitation of resource potential and conforms to an energy transition in alignment with global trends. These trends encompass intra-fossil substitution, particularly the displacement of coal and liquid fuels by natural gas, the escalating integration of unconventional renewable energies into the secondary matrix (primarily wind and solar, with considerable potential for hydropower), a reduction in the energy intensity rate (achieving higher output with reduced energy consumption), augmented electrification of final consumption (encompassing electric vehicles and other applications), the implementation of smart grids (involving the digitization of electrical networks), and the diminishment of the carbon footprint per unit of energy utilized in consumer-demanded goods and services. In certain instances, adherence to these principles becomes imperative for exporting to select nations without incurring penalties related to environmental concerns. In this context, the Argentine government has, in recent years, promulgated diverse regulatory measures, policies, and initiatives aimed at fostering and expediting the ongoing energy transition, with the principal endeavors outlined subsequently.

In this line of thought, we consider a sustainable energy transition demands foresight and commitment. In this chapter we propose the development of a long-term vision that must transcend political cycles and short-term fluctuations. This requires policy continuity, robust regulatory frameworks, and the establishment of clear milestones to guide Argentina's trajectory towards a sustainable energy future. Successful energy transition would require a holistic and coordinated effort, involving multi-stakeholder collaboration and a commitment to long-term sustainability. Regular monitoring, assessment, and adjustments to policies and strategies will be crucial for the effectiveness of the decarbonization scenario.

A one-size-fits-all approach is clearly inadequate, and therefore, it is important to formulate strategies that align with Argentina's unique socio-economic and geographical characteristics, integrating local expertise, engaging communities, and fostering public–private partnerships to ensure the success and sustainability of the proposed energy transition strategy.

The pillars of the Argentine energy transition are rooted in the concerted efforts towards transitioning to massive electric power generation sourced from renewable and clean energy sources. This shift marks a pivotal move away from reliance on fossil fuels, aligning with global trends towards sustainability and environmental preservation. Concurrently, while the transition to renewables gains momentum, natural gas assumes a crucial transitional role. Positioned as a bridge fuel, natural gas provides a reliable and less carbon-intensive alternative, facilitating a smoother transition towards a cleaner energy landscape.

Moreover, the Argentinean energy transition necessitates a multifaceted approach, requiring various complementary actions to propel the nation towards its energy goals. These actions encompass not only the expansion of renewable energy infrastructure but also the modernization and optimization of existing energy systems. Investment in research and development, as well as the implementation of innovative technologies, plays a pivotal role in enhancing energy efficiency and driving sustainable growth.

Furthermore, fostering a supportive regulatory framework and incentivizing private sector participation are essential components of the transition process. Encouraging investment in renewable energy projects, streamlining permitting processes, and providing financial incentives can accelerate the adoption of clean energy technologies. Additionally, fostering partnerships and collaboration between government entities, private enterprises, and civil society organizations can foster knowledge-sharing and innovation, further propelling the energy transition forward.

In summary, the Argentine energy transition represents a multifaceted and dynamic process aimed at reshaping the nation's energy landscape towards sustainability and resilience. By embracing renewable energy sources, leveraging natural gas as a transitional fuel, and implementing complementary actions across multiple fronts, Argentina can position itself as a leader in the global transition towards a cleaner and more sustainable energy future.

References

Allub L et al (2024) Report on economy and development: renewed energies: a just energy transition for sustainable development. CAF—Latin American and the Caribbean Development Bank, available at https://scioteca.caf.com/handle/123456789/2250

Ameli N, Brandt N (2015) Determinants of households' investment in energy efficiency and renewables: evidence from the OECD survey on household environmental behaviour and attitudes. Environ Res Lett 10(4):044015

Audoly R, Vogt-Schilb A, Guivarch C, Pfeiffer A (2018) Pathways toward zero-carbon electricity required for climate stabilization. Appl Energy 225(November 2017):884–901

Bataille C (2020) Physical and policy pathways to net-zero emissions industry. WIRES Wiley Interdiscip Rev Clim Change 11(e633):1–20

Bataille C et al (2016) The need for national deep decarbonization pathways for effective climate policy. Clim Policy 16(sup1):S7–S26

Belfiori E, Rezai A (2024) Implicit carbon prices: making do with the taxes we have. J Environ Econ Manag 125:102950

Bento AM, Goulder LH, Jacobsen MR, von Haefen RH (2009) Distributional and efficiency impacts of increased US gasoline taxes. Am Econ Rev 99(3):667–99

Calatayud A, Sánchez González S, Bedoya-Maya F, Giraldez Zúñiga F, Márquez JM (2021) Urban Road Congestion in Latin America and the Caribbean: Characteristics, Costs, and Mitigation

Blackman A, Li Z, Liu AA (2018) Efficacy of command-and-control and market-based environmental regulation in developing countries. Ann Rev Resour Econ 10:381–404

Burtraw D, Sweeney R, Walls M (2009) The incidence of U.S. climate policy: alternative uses of revenues from a cap-and-trade auction. Nat Tax J 62(3):497–518

Cont W, Navajas F (2019) Subsidios a los servicios de infraestructura en América Latina y el Caribe: direcciones de reforma. DC. Unpublished mimeo, Inter-American Development Bank, Washington

Cont W, Romero C, Lleras G, Unda R, Celani M, Gartner A, Cepelli L, Zipitría L, Besfamille M, Figueroa N, López Azumendi S, Fischer R (2021) IDEAL 2021: The impact of digitalization to reduce gaps and improve infrastructure services. Caracas: CAF. http://scioteca.caf.com/handle/123456789/1762

Cont W, Belfiori E, Rodríguez Pardina M, Rojas D, Fernández S, Bonifaz JL et al (2022) IDEAL 2022: energy, water, and health for a better environment. Retrieved from https://scioteca.caf.com/handle/123456789/2017

Clarke L et al (2014) Assessing transformation pathways. In: Climate change 2014: mitigation of climate change. Contribution of working group III to the fifth assessment report of the intergovernmental panel on climate change. In: Edenhofer O et al (eds) Cambridge University Press. Cambridge, UK and New York, NY, USA, pp 413–510

Cole W, Frazier AW, Augustine C (2021) Cost projections for utility-scale battery storage: 2021 update (No. NREL/TP-6A20–79236). National Renewable Energy Lab (NREL), Golden, CO (United States)

Cramton P, Ockenfels A, Tirole J (2020) Policy brief—translating the collective climate goal into a common climate commitment. Rev Environ Econ Policy

DeAngelo J, Azevedo I, Bistline J, Clarke L, Luderer G, Byers E, Davis SJ (2021) Energy systems in scenarios at net-zero CO_2 emissions. Nat Commun 12(1):6096

Deloitte–Enel (2022) Hoja de ruta para la transición Energética. 2030–2050. Taller III: Resultado Finales y Recomendaciones. 4 de Octubre 2022

Deloitte–Enel (2023) Actualización de la Hoja de ruta de Transición Energética en Argentina: Un modelo energético sostenible para Argentina al 2050, Febrero 2023

De Vries LJ, Verzijlbergh RA (2018) How renewable energy is reshaping Europe's electricity market design. Econ Energy Environ Policy 7(2):31–50

Di Bella G, Norton L, Ntamatungiro J, Ogawa S, Samaké I, Santoro M (2015) Subsidios a la energía en América Latina y el Caribe: Balance y desafíos en relación a las políticas. Documentos de trabajo, Fondo Monetario Internacional 15(30):1

Egerer J, Rosellón J, Schill W-P (2015) Power system transformation toward renewables: an evaluation of regulatory approaches for network expansion. Energy J Int Assoc Energy Econ 36(4)

Feng K, Hubacek K, Liu Y, Marchán E, Vogt-Schilb A (2018) Efectos distributivos de los impuestos a la energía y de la eliminación de los subsidios energéticos en América Latina y el Caribe (No. IDB-WP-947). IDB Working Paper Series

Gao X, Zhou S (2022) Solar adoption inequality in the US: trend, magnitude, and solar justice policies. Energy Policy 169:113163

Gonzales LE, Ito K, Reguant M (2022) The dynamic impact of market integration: evidence from renewable energy expansion in Chile (No. w30016). National Bureau of Economic Research

Graham K, Knittel CR (2024) Assessing the distribution of employment vulnerability to the energy transition using employment carbon footprints. Proc Natl Acad Sci 121(7):e2314773121

Guo X, Lin F, Sun X (2021) Can environmental regulations promote greenhouse gas abatement in OECD countries? Command-and-control vs. market-based policies. Sustainability 13(12):6913

Gutiérrez-Meave R, Rosellón J, Sarmiento L (2021) The effect of changing marginal-cost to physical-order dispatch in the power sector, DIW Discussion Papers Num. 1955, 2021.
Hancevic P (2017) A dynamic approach to environmental compliance decisions in US electricity market: the acid rain program revisited. Energy Policy 106:129–137
Hancevic P, Cont W, Navajas F (2016) Energy populism and household welfare. Energy Econ 56:464–474
Hancevic P, Núñez H, Rosellón J (2017) Distributed photovoltaic power generation: Possibilities, benefits, and challenges for a widespread application in the Mexican residential sector. Energy Policy 110:478–489
Hancevic P, Núñez H, Rosellón J (2022a) Electricity tariff rebalancing in emerging countries: the efficiency-equity tradeoff and its impact on photovoltaic distributed generation. Energy J 43(4)
Hancevic P, García B, Bejarano H, Sandoval H (2025) Incentivizing Solar: Experimental Insights for Mexico's Renewable Energy Policy, CIDE Working Paper
Hancevic P, Núñez H, Rosellón J (2022b) Mexico's energy prospects: gains from renewable sources over a fossil fuel-dominated environment. Econ Energy Environ Policy 11(2)
Hancevic P, Núñez H, Rosellón J (2023) El sector energético en América Latina y el Caribe: oportunidades y desafíos del cambio climático. CAF-Banco de Desarrollo de América Latina, Policy Paper 18
IEA (2023) Latin America energy outlook 2023. https://www.iea.org/reports/latin-america-energy-outlook-2023
IMF (2023) Fossil Fuel Subsidies by Country and Fuel Database (2023), available at https://www.imf.org/-/media/Files/Topics/energy-subsidies/EXTERNALfuelsubsidiestemplate2023new.ashx
IRENA (2020a) Renewable power generation costs in 2020, International Renewable Energy Agency
IRENA (2020b) Power system organizational structures for the renewable energy era, 36. International Renewable Energy Agency
IRENA (2020c) Green bonds—renewable energy finance brief 03. January 16. International Renewable Energy Agency
IRENA (2020d) Renewable power generation costs in 2019, 160. International Renewable Energy Agency
Kaika D, Zervas E (2013) The environmental Kuznets curve (EKC) theory-Part A: concept, causes and the CO_2 emissions case. Energy Policy 62:1392–1402. https://doi.org/10.1016/j.enpol.2013.07.131
Levy S (2020) Superemos juntos la emergencia. https://www.nexos.com.mx/?p=47405
Liang Q, Wei Y (2012) Distributional impacts of taxing carbon in China: results from the CEEPA model. Appl Energy 92 (abril):545–551
López Soto D, Mejdalani A, Nogales A, Tolmasquim M (2019) Avances en el diseño de políticas y marcos regulatorios para las energías renovables en América Latina y el Caribe para la generación distribuida y a escala de la red de distribución eléctrica. Inter American Development Bank
Markandya A, Wilkinson P (2007) Electricity generation and health. Lancet 370(9591):970–990
Martínez-Cruz AL, Núñez HM (2021) Tension in Mexico's energy transition: are urban residential consumers in Aguascalientes willing to pay for renewable energy and green jobs? Energy Policy 150:112145
Ortner A, Totschnig G (2019) The future relevance of electricity balancing markets in Europe-A 2030 case study. Energ Strat Rev 24:111–120
OLADE (2023). Reservas de hidrocarburos. SieLac, available at: https://sielac.olade.org/WebForms/Reportes/ReporteDato3.aspx?oc=58&or=670&ss=2&v=1
Pai S, Emmerling J, Drouet L, Zerriffi H, Jewell J (2021) Meeting well-below 2°C target would increase energy sector jobs globally. One Earth 4(7):1026–1036
Parry I, Williams R (2010) What are the costs of meeting distributional objectives for climate policy? B.E. J Econ Anal Policy 10(2):1935–1682

Parry I, Heine D, Lis E, Li S (2014) Obtener precios de energía correctos: Del principio a la práctica. Fondo Monetario Internacional

Parry, I., Black, S., and Vernon, N. (2023) Measuring Fossil Fuel Subsidies-A Global and Country View. Data for Greener World: A Guide for Practitioners and Policymakers, 95-112.

Quiros-Tortos J, Victor-Gallardo L, Ochoa L (2019) Electric vehicles in Latin America: slowly but surely toward a clean transport. IEEE Electrication Mag 7(2):22–32

Rapson D, Muehlegger E (2021) The economics of electric vehicles. NBER Working paper 29093, July

Ritz RA (2022) Global carbon price asymmetry. J Environ Econ Manag 102687

Rockström J et al (2017) A Roadmap for Rapid Decarbonization. Science 355(6331):1–2

Rosemberg A (2017) Strengthening just transition policies in international climate governance. https://stanleycenter.org/publications/pab/RosembergPABStrengtheningJustTransition417.pdf

Sepulveda NA, Jenkins JD, de Sisternes FJ, Lester RK (2018) The role of firm low-carbon electricity resources in deep decarbonization of power generation. Joule 2(11):2403–2420

Saget C, Vogt-Schilb A, Luu T (2020) Jobs in a net-zero emissions future in Latin America and the Caribbean

Sanguinetti P, Moncarz P, Vaillant M, Allub L, Juncosa F, Barril D, Cont W, Lalanne Á (2021) RED 2021: pathways to integration: trade facilitation, infrastructure and global value chains. CAF, Caracas. http://scioteca.caf.com/handle/123456789/1823

Secretary of Energy (2022) Plan Nacional de Expansión del Transporte Eléctrico 2035

Solano-Rodríguez B et al (2021) Implications of climate targets on oil production and fiscal revenues in Latin America and the Caribbean. Energy Clim Change 2(May):100037

Symons EJ, Speck S, Proops JLR (2002) The distributional effects of carbon and energy taxes: the cases of France, Spain, Italy, Germany and UK. Eur Environ 12(4):203–212

Tong D et al (2019) Committed emissions from existing energy infrastructure jeopardize 1.5 °C climate target. Nature 572(7769):373–377

Undersecretary of Energy Planning (2023) Lineamientos y Escenarios para la Transición Energética a 2050. Mayo 2023

Welsby D, Rodriguez BS, Pye S, Vogt-schilb A (2021) High and dry: stranded natural gas reserves and fiscal revenues in Latin America and the Caribbean. Banco Interamericano de Desarollo, Documento de Trabajo 01197

Williams JH et al (2012) The technology path to deep greenhouse gas emissions cuts by 2050: The pivotal role of electricity. Science (New York, N.Y.) 335(6064):53–59

Wood Mackenzie (2024) Latin America levelized cost of electricity 2023 (LCOE), Commodity Market Report, February 2024

World Bank (2017) Charting the diffusion of power sector reforms across the developing world. Policy Research Paper no. 8235. World Bank, Washington, D.C. License: CC BY 3.0 IGO. http://hdl.handle.net/10986/28853

Part III
Human-Centered Climate Actions: Empowering Individuals and Younger Generations

Chapter 5
The Role of Behavioral Sciences in the Communication of Climate Change: *Insights and Opportunities for Argentina*

Ailin Tomio and Joaquin Navajas

Abstract Like most countries in the region, Argentina faces the challenge of promoting climate action to mitigate greenhouse emissions and adapt to changes in environmental patterns. Traditional science communication methods, which often rely on statistical data and abstract concepts, are deemed insufficient for driving behavioral change. In this chapter, we leverage insights from behavioral science to emphasize the importance of addressing psychological determinants of pro-environmental behavior, such as environmental self-identity, perceived behavioral plasticity, and outcome efficacy. Additionally, we highlight the role of social factors, particularly political polarization, in shaping public attitudes and actions toward climate change. This non-systematic review identifies a gap between climate acceptance and climate action in the literature, suggesting that interventions should not only focus on communications but also on enhancing convenience and reducing the friction associated with sustainable choices. We highlight relevant frameworks to understand goal conflicts influencing environmental behavior, with recommendations for promoting pro-environmental actions by altering perceived outcomes and strengthening normative goals. Practical recommendations for developing behaviorally informed campaigns to promote climate action in Argentina are provided. These include using simple and clear messages, tailored persuasive appeals, culturally relevant storytelling, audience segmentation, and policy interventions that leverage default options and extrinsic incentives. We emphasize the need for further

A. Tomio (✉)
Mind, Behavior and Development Unit, World Bank, Washington D.C., USA
e-mail: atomio@udesa.edu.ar

Departamento de Educación, Universidad de San Andrés, Buenos Aires, Argentina

J. Navajas
Laboratorio de Neurociencia, Escuela de Negocios, Universidad Torcuato Di Tella, Buenos Aires, Argentina
e-mail: joaquin.navajas@utdt.edu

Consejo Nacional de Investigaciones Científicas y Técnicas (CONICET), Buenos Aires, Argentina

E. Belfiori and G. Gertner (eds.), *Climate Change and Economic Growth*, SpringerBriefs in Latin American Studies, https://doi.org/10.1007/978-3-031-99637-5_5

research on the behavioral determinants of climate action in Argentina to inform effective communication strategies and policy interventions.

Keywords Behavioral sciences · Climate change · Behavioral public policy · Science communication · Political polarization · Recommendations · Argentina

5.1 Introduction

Despite posing a significant social challenge in the twenty-first century, public involvement in addressing climate change remains notably low worldwide. Although general awareness of the issue has increased in the past decades, a considerable portion of the population perceives climate change as a non-urgent and psychologically distant risk characterized by spatial, temporal, and social remoteness (van der Linden et al. 2015). This prevailing perception has resulted in a delay in public decision-making concerning both mitigation and adaptation responses to the climate crisis. In this chapter, we present an outline of recent findings from the field of the behavioral sciences to communicate climate change issues and promote positive behavior change.

The consequences of climate change already impact human health and life. These include rising sea levels, more frequent and severe natural disasters such as hurricanes, droughts, floods, changes in precipitation patterns, loss of biodiversity and heat-driven mortality (Intergovernmental Panel on Climate Change 2018). When talking about climate change, the consumption choices of individuals related to transportation, food, and housing contribute directly or indirectly to an essential percentage of global greenhouse gas (GHG) emissions (IPCC 2022). For example, a study based on US markets found that if six behaviors were adopted by 5–10% of the population, they could collectively reduce current national emissions by an estimated 464 million metric tons of carbon dioxide per year (Heller et al. 2021). These behaviors are purchasing an electric vehicle (EV), reducing air travel, eating a plant-rich diet, purchasing carbon offsets, reducing food waste, and purchasing green energy. In developing countries, acknowledging climate change becomes key for governments and populations to prepare adaptation strategies and increase their resilience to climate change consequences.

Behaviors such as reforestation, using clean energy for cooking, cooling or heating, and preparedness for risky situations can effectively address sustainable development priorities in disadvantaged territories. Argentina, for example, is already facing losses caused by droughts and floods. By 2050, drought losses could account for 4% of Argentina's GDP. Furthermore, floods cause up to $1.4 billion average annual asset losses and $4 billion in welfare losses (World Bank Group 2022). The country has been ranked in the 84th position according to the Climate Vulnerability Index 2023 (ND-GAIN Country Index 2023) and thus needs to commit to transition to a low-carbon economy. In doing so, it can significantly increase its economic growth and the well-being of its people. In addition, Argentina ranks 22nd among greenhouse gas emitters worldwide (World Resources Institute 2022). Compared to its neighboring countries, Argentina is the second country with the most CO_2 emissions per capita after Chile (Global Carbon Budget 2023). Emissions come mostly from energy (51%, including 16% from energy industries and 13% from transport), followed by agriculture, livestock, forestry, and other land use (39%), industry (6%), and waste (4%) (Ministerio de Ambiente y Desarrollo Sostenible de Argentina 2021). Consequently, shifting climate behaviors in Argentina has become a pressing issue. Household transition to clean energy appliances, shifting transport practices to use electric vehicles or sustainable transport, and reducing meat consumption are a few behaviors that can help the country achieve its reduction goals.

Policies that use behavioral science principles to encourage individual action while preserving choice are effective for shifting behavior and present a complementary path to more traditional policy tools (Sunstein 2016a, b; Heller et al. 2021). However, there is agreement among scientists that climate change is a collective action problem requiring collective action solutions (Fengler 2022). Individuals can significantly mitigate climate change through citizen and collective behavior (Steg 2023), for example, by joining climate strikes or voting for political parties with ambitious climate goals. They can also impact the policies and actions of the organizations they are a part of, encouraging them to adopt sustainable practices and reduce their greenhouse gas emissions. Furthermore, individuals can support technology, policy, and system changes to reduce greenhouse gas emissions, thus contributing to the overall effort to combat climate change (Stern 2016; Wolske and Stern 2018).

Addressing the key factors influencing sustainable choices and eliminating behavioral barriers is crucial to promoting collective climate action. Although several studies suggest that targeting awareness and beliefs related to climate change can have a moderate impact on their behaviors (Steg 2023), there is an apparent disparity between public communication and the lack of actualized behavioral change observed in the general public (Whitmarsh 2008). While public polls often indicate that people express general awareness and concern (e.g., Environics International Environmental Monitor 2000; GlobeScan 2006), individuals remain reluctant to take personal action. In terms of concerns, Argentina does not lag behind. In a survey conducted in 2018 across 26 nations, most countries ranked climate change as the most important global threat. In Argentina, 73% of respondents considered climate change a significant threat to the country (Poushter and Huang 2019). However, similar increases in environmental behavior have not accompanied this shift in attitudes, which underscores the need to understand the different factors that mediate between beliefs, attitudes, and behaviors.

Unfortunately, behavioral studies addressing climate change actions in Argentina are scant. For example, in the Patagonian city of Trelew, the World Bank implemented a qualitative diagnosis and designed a randomized controlled trial to test the effect of informational flyers and magnetic calendars on recycling behaviors. Households that received the interventions had nearly double (31%) source segregation rates relative to the control group (17%). The simplified information condition was most effective, which told residents when to dispose of recyclable waste and included a table with relevant recyclable and residual or organic waste items (Pegels et al. 2022). Another example is the case of two field studies that tested the effects of a charge for single-use plastic bags in Buenos Aires City. Jakovcevic et al. (2014) demonstrated a substantial increase in consumers' use of reusable bags after implementing a plastic bag charge in participating supermarkets. This increase was notably greater compared to control supermarkets where the charge was absent or subsequently introduced. The observed effects were further amplified two months post-implementation. The second study, conducted six months later by the same researchers, revealed factors influencing policy support and reusable bag use. Supporters of the policy often emphasized environmental benefits, whereas opponents primarily declared concerns related to financial costs. Additionally, most consumers indicated using their own bags out of a desire to protect the environment, implying that behavioral changes were largely driven by intrinsic motivations rather than extrinsic factors. Cases like this should be replicated and adapted to local communities, targeting the key behaviors that can help the country achieve its climate goals.

This chapter provides an overview of the state of the art in the behavioral sciences and sustainability to help researchers and policymakers discover opportunities for future regional studies. The chapter is structured as follows: Sect. 5.2 presents the role of behavioral science in bridging the gap between scientific insights and their application in society. Section 5.3 highlights the individual determinants of pro-environmental behavior. Section 5.4 focuses on the influence of political polarization on science attitudes and behaviors and highlights the need for effective communication strategies. Section 5.5 delves into the relationship between climate change acceptance and climate action, providing a deeper dive into the physical drivers of sustainable behaviors. Finally, Sects. 5.6 and 5.7 summarize the learnings from behavioral science and provide some practical recommendations for policymakers.

5.2 The Problem with Science Communication

Climate change has usually been studied and thus communicated in statistical terms (e.g., by showing long-term trends in temperature) (van der Linden et al. 2015). However, abstract, descriptive, and analytical formats may not be the best way to reach large audiences. Decades of research in social, cognitive, and clinical psychology have shown that people do not process uncertain information in a logical and analytical manner (Marx et al. 2007), but often rely on intuitive, experiential, automatic, and affective (emotional) processing (LeDoux 1989; Marx et al. 2007). The persistent lack of public engagement with these issues may result from communications that represent psychologically distant concepts and fail to convince people that their behavior can contribute to the outcome.

The degree to which an individual's behavior can be changed is crucial in assessing the potential for mitigating negative impacts. Behavioral plasticity refers to the ability of an individual to modify their behavior in response to changing circumstances or stimuli. For example, if someone is trying to reduce their carbon footprint, their ability to switch to more sustainable transportation options or reduce their energy consumption would be an example of behavioral plasticity. By understanding whether and how individuals can modify their behavior, behavioral scientists can better assess the potential for mitigating negative impacts (Dietz et al. 2009).

The malleability of human behavior is related to the individual's motivation to change their actions, their possession of the necessary skills, and the existence of structural, contextual, social, and economic barriers that impede behavior change (Guagnano et al. 1995; Black et al. 1985). These barriers vary significantly among households, types of behavior, and geographical contexts. Needless to say, poor households in developing countries tend to exhibit lower behavioral plasticity as their ability to change is constrained by their dependency on natural capital to earn money and their lack of access to sustainable alternatives.

Nielsen et al. (2024) provide a framework that categorizes the determinants of sustainable behaviors at four levels: systemic, physical, social or individual. The interplay between these factors results in different behavioral outcomes. The system level observes the interaction between government institutions and organizations and their legislation and market dynamics. The physical level considers access to relevant public and commercial services as well as economic determinants. The social category considers community behaviors, social habits, and perceptions. Finally, the individual level includes personal motivations, values, and levels of knowledge and awareness. For example, an individual's choice of a plastic product might be determined by the products offered in the market (system), how accessible they are in nearby stores (physical), the daily needs of their family or group (social), and their personal preferences (individual). This chapter will focus on individual (psychological determinants), social (polarization), and physical (convenience) factors that can alter how people behave toward climate change.

5.3 The Individual Level: Behavioral Determinants of Pro-Environmental Behaviors

Scientific communications on climate actions can impact behavior change to the extent they can alter how people perceive their role in mitigating or adapting to negative climate effects. For example, high perceived behavioral plasticity correlates positively with environmental self-identity. Environmental self-identity refers to the extent to which people consider themselves pro-environmental (Abrahamse 2019). People try to be consistent between their thoughts and actions and thus are motivated to act in line with their self-image (Cialdini 2001). Environmental self-identity can therefore impact various environmental behaviors across diverse contexts. Additionally, individuals derive their identities from past behaviors, indicating that environmental self-identity is malleable and responsive to change.

A study of over 7000 individuals on climate-related behaviors and opinions found that environmental identity was positively associated with the perceived behavioral plasticity of different household behaviors and policy support (Kukowski et al. 2023).

This suggests that communications targeting awareness creation should address psychological determinants such as environmental self-identity to promote positive behavior change.

Cost-effective interventions based on behavioral science literature have produced positive results in increasing awareness and support on these issues (Nisa et al. 2019; Mertens et al. 2022; van Valkengoed et al. 2022). Results from a second-order meta-analysis (Bergquist et al. 2017), including 10 meta-analyses and 430 primary studies, show that pro-environmental behaviors can be increased by 2 to 12 percentage points when using behavioral strategies to persuade action. However, the effectiveness of a specific intervention is dependent on the key determinants (variables that act as barriers or levers of the behavior) and characteristics of the targeted behavior. Within the behavior change literature, it is typically assumed that interventions do not directly change behavior. Rather, interventions target one or more behavioral determinants, which in turn results in behavior change.

van Valkengoed et al. (2022) reviewed relevant theoretical frameworks and literature and proposed a classification system that links six types of interventions to 13 determinants of environmental behavior. Table 5.1 outlines the main determinants selected by the authors.

These determinants can be very useful when trying to design an effective message to target specific behaviors. For example, a recent large-scale study across 63 countries evaluated the efficacy of different expert-crowdsourced messages on multiple outcome variables such as belief in climate change, support for climate change mitigation policy, willingness to share climate mitigation information on social media, and tree planting behavior (Vlasceanu et al. 2024). The study found that interventions varied in effectiveness across outcomes. Reducing psychological distance proved effective in increasing belief in climate change (2.3% effect size), while writing a letter to a future generation was most impactful in garnering support for climate policies (2.6% effect size). The largest gains were observed in the negative emotion induction condition related to willingness to share information (12.1%). Notably, the effects of each intervention varied depending on people's initial climate beliefs. Most interventions were relatively ineffective among those low in initial belief (i.e., climate skeptics or those with low environmental self-identity). The main exception was writing a letter to a member of the future generation, which worked across most of the spectrum of initial belief. The findings of this study underscore the importance of tailoring interventions based on specific outcomes and target populations, urging a locally customized and adaptable approach in the field.

When focusing on changing behaviors, the effects of interventions are still smaller. In the previously mentioned mega study, a tree planting task (a cognitively demanding task) was used to measure objective pro-environmental behavior. Interestingly, more than half of the interventions decreased the number of completed tasks across all levels of initial belief (Vlasceanu and Coman 2018). One of the reasons for these small effects may be related to the fact that many environmental behaviors generate a conflict between convenience or individual gain versus normative and collective goals (Steg et al. 2014). These dimensions will be further explained in the following sections.

Table 5.1 Main determinants of pro-environmental behavior according to van Valkengoed et al. (2022)

Determinants	Description
Attitudes toward the behavior	The degree to which a person positively or negatively evaluates a particular behavior based on the beliefs and evaluation of the costs and benefits of behavior. The more positive the attitude, the more likely they are to engage in that behavior
Self-efficacy	How capable people feel of implementing a specific action. The more people feel they can implement a particular behavior, the more likely they are to do so
Injunctive norms toward the behavior	The extent to which people believe a behavior is commonly approved or disapproved of by people or groups they identify with. The more people perceive an injunctive norm to act pro-environmentally, the more likely they are to comply with it
Descriptive norms toward the behavior	How much people believe others engage in a behavior and whether the behavior is normative. When people think that many others engage in a behavior, they are more likely to follow
Problem awareness	Awareness of the relationship between actions and climate issues. Higher problem awareness correlates positively with more engagement in sustainable behaviors
Ascription of responsibility	The extent to which people feel responsible for the (negative) environmental consequences of their actions. The more responsible people feel, the more likely they are to engage in pro-environmental behavior
Outcome efficacy	How much people perceive their behavior as effective in resolving environmental problems. The more people perceive their behavior as effective in reducing environmental problems, the more likely people are to engage in that action
Environmental self-identity	The extent to which people think of themselves as pro-environmental. As indicated above, people are motivated to act in line with their self-image
Risk perception	The more risk people perceive, the more likely they are to engage in behavior to decrease that risk, such as pro-environmental behavior

5.4 The Social Determinants: Communicating Science in Polarized Societies

At the social level, people are affected by the political context in which they are embedded. Encouraging individuals to engage in behavioral actions to reduce the impact of climate change becomes more challenging in societies with strong political polarization (Cole et al. 2023). Political polarization can be expressed in diverse forms, although not all manifestations are necessarily bad for democracy. For instance, issue polarization, characterized by people having divergent views on policymaking, may foster rational debates of ideas between opposing political platforms. However, amidst these nuanced variations, political polarization can also be manifested in less positive manners and threaten societal cohesion. Affective polarization emerges when individuals not only hold contrasting political views but, more critically, undermine and neglect the validity of disagreeing opinions. In such a context, citizens may find themselves retreating into ideological echo chambers, limiting interactions with fellow citizens and reducing the exchange of diverse perspectives essential for informed decision-making. This divisive trend has already noticeably undermined cooperation in economic markets (McConnell et al. 2018), strained interpersonal relationships (Keith Chen and Rohla 2018), and played an important role in the residential segregation of communities.

The intricate intersection of political polarization and the challenge of addressing climate change is evident, as divisions in support of public policies emerge conspicuously at both elite (Fisher et al. 2013) and mass levels (Falkenberg et al. 2022). While most existing data primarily emanates from the USA, there is a discernible trend indicating a global alignment between political polarization and the level of support for climate action initiatives (Birch 2020). The manifestation of this alignment transcends geographical boundaries and represents a multifaceted phenomenon with far-reaching implications, warranting a comprehensive exploration of its implications on public discourse (Feldman and Hart 2018), policy formulation (Hart and Nisbet 2012), and societal resilience to the challenges posed by a changing climate (McEvoy et al. 2013).

Argentina operates under a representative democratic system, with a federal structure that includes a president elected by popular vote for a four-year term. Argentina has been historically marked by political polarization, with sharp divisions between Peronist and non-Peronist factions dominating much of the political landscape. This polarization is evident in the alternating power dynamics between the two major political parties, the Peronist Justicialist Party and various opposition groups. Additionally, ideological differences and contentious issues, including whether the government should continue funding public healthcare or if the country should migrate to a primarily private model similar to the one existing in the USA.

In contrast to the situation in the USA, where those advocating for less government intervention in health care are typically referred to as "conservatives," in Argentina, individuals supporting similar views are often labeled as "liberals." These Argentine "liberals" align more closely with anarcho-libertarian perspectives found in the USA, emphasizing minimal government involvement and maximum individual freedom. This nuanced understanding of socio-cultural differences, rooted in the history of the Peronist movement, highlights the complexity of political ideology in Argentina, where the traditional liberal-conservative spectrum fails to capture the diversity of viewpoints fully.

In Argentina, although there is presently limited evidence of polarization on climate action across the political spectrum, the recent success of a political platform that explicitly denies climate change poses a potential hazard (Ramos Muñoz 2023). Drawing comparisons with the politicization and polarization of backing for scientific advice during the COVID-19 pandemic, this situation raises concerns about Argentina's capacity to cohesively respond to climate change. For instance, past research has illustrated that support for COVID-19 prevention policies in Argentina was consistent with a tribal partisan model (Van Bavel and Pereira 2018). Individual beliefs about the severity of the pandemic played a negligible role in shaping policy support (Freira et al. 2021), with partisanship emerging as the primary factor. Similarly, partisan affiliations predicted trust in specific vaccine brands (Fumagalli et al. 2023), perceptions of vaccination campaign effectiveness (Fumagalli et al. 2023), and susceptibility to COVID-19-related disinformation (Espina Mairal et al. 2024).

However, in the midst of a polarized context, one might ask: what measures can be undertaken to effectively communicate scientific recommendations? To answer this question, one may focus on the integration of behavioral science into evidence-based policymaking during public health emergencies like the COVID-19 crisis, which may provide a pertinent framework for addressing analogous challenges posed by climate change. In this context, a recent study authored by a global consortium of researchers has found that bipartisan support for specific policies is essential in reducing polarization on scientific recommendations (Flores et al. 2022). This evidence, gleaned from an evaluation of evidence-based policymaking during the pandemic, is extremely relevant for guiding societal responses in the not-so-unlikely scenario of a major climate emergency.

5.5 The Physical Level: The Gap Between Climate Acceptance and Climate Action

Even when climate change is recognized as an urgent global threat by leaders across the whole political spectrum, there might still be a gap between awareness and action. Indeed, findings show that people generally acknowledge climate change's existence and negative consequences (Capstick et al. 2015; Leiserowitz et al. 2021). This may explain why efforts to increase climate change perceptions have relatively weak effects in several studies (Rode et al. 2021). This is the case for Argentina, where 85% of respondents believe climate change impact their everyday lives, and 83% favor stricter government measures to foster the adoption of climate-friendly behavior. Most importantly, 67% say climate change affects their income or livelihood (Griffith-Jones and Tyson 2013). However, these concerns are not reflected in actual behaviors. For example, according to the Ministry of Health and Environment, in 2018, Argentina produced around 0.86 kg of waste per person per day, and only 10% was recycled, with the remainder being landfilled or improperly disposed of in dumpsites (Hotspot 2021). In terms of energy use, residential energy consumption accounts for 26% of the total final energy consumption in the country, representing the third sector that consumes the most after industry and transport (IEA Data Services 2022).

Behavioral science studies have shown that beyond awareness and beliefs, interventions should target convenience. Interventions of this sort focus on nudging behavior by removing the hassle of acting sustainably and improving access to the physical elements that can promote the desired behavior. For example, a recent study in China showed that setting "no-cutlery" as the default option in Alibaba's mobile food delivery platform increased the share of no-cutlery orders by 648% (He et al. 2024). In the USA, an analysis of 13 metropolitan areas showed that when bike lanes are designed to protect cyclists from motor traffic, residents are more likely to cycle (Buehler et al. 2009). Even reducing the friction on desired behaviors by a small amount can significantly impact habits. A study conducted in an office setting found that the uptake of recycling doubled when signs were placed right next to trash bins rather than at a short distance of approximately 4 m away (Austin et al. 1993).

Although reducing the costs of pro-environmental choices is important when they present frictions, it may not result in sustained pro-environmental actions, even when the costs are reduced. One possible reason may be due to effects that do not generalize to other contexts or behaviors and may not persist when the intervention is removed (Mols et al. 2015). Steg and colleagues (2014) present a theoretically valuable framework for tackling barriers to sustainable behaviors. According to the Integrated Framework for Encouraging Pro-environmental Behavior (IFEP), conflicting goals often shape environmental behavior (Steg et al. 2014). Three types of goals—hedonic, gain, and normative—rule individuals' environmental behavior in specific situations. Hedonic goals prioritize improving feelings, gain goals focus on personal resource changes, and normative goals center around appropriateness and moral considerations. The prominence of a particular goal in a situation, termed the "goal frame," significantly influences cognitive processes and decision-making, with other goals either reinforcing or diminishing the strength of the focal goal. These goal dynamics occur subconsciously, influenced by individual dispositions and environmental cues, contributing to a nuanced understanding of environmental decision-making.

Based on this, the authors propose two fundamental strategies for promoting pro-environmental actions in the face of goal conflicts. The first strategy involves altering the behavior's actual or perceived outcomes to mitigate the conflict between hedonic or gain goals versus normative goals. This entails reducing the perceived costs of pro-environmental actions while increasing their perceived benefits. This strategy is commonly employed through information campaigns, pricing policies, or physical modifications. The second strategy focuses on strengthening normative goals to diminish the relative influence of hedonic and gain goals. By emphasizing the environmental consequences of behavioral choices, this approach encourages individuals to engage in pro-environmental actions driven by a sense of moral duty, even if these actions are less convenient or more costly.

In line with Steg and her colleagues, Newell and Moss (2021) suggest that communicating moral arguments to persuade people to behave more sustainably can increase intervention effectiveness. According to these authors, messages should contain the following elements: reference each person's duties (injunctive norms), demonstrate that individual actions are effective (self-efficacy), and help people define their fair share. Social norms are another relevant driver of behavior. Imitating the behavior of your group or the majority reduces the cost of decision-making (van der Linden et al. 2015). In the particular context of climate change, people are more likely to act if their social references are doing the same action as they perceive group effort to be more effective since climate change is a collective issue (Newell and Moss 2021).

Beyond relying on the norm, van der Linden et al. (2015) recommend appealing to the experiential processing system, highlighting relevant personal experiences through recall, scenarios, and storytelling. Individuals vary in their inclination for cognitive effort, often acting as "cognitive misers" by resorting to mental shortcuts, especially under the umbrella of bounded rationality, where decisions are made within cognitive constraints, task environments, and available information. Mental shortcuts can be intentional, relying on rules of thumb like deference to experts or consensus, or automatic, manifesting as biases like availability, representativeness, and anchoring and adjustment (Maibach et al. 2023). Emotions and past experiences influence the use of mental shortcuts and the decision-making processes.

In science communication, reliance on mental shortcuts poses challenges, as overly complex information may lead to disengagement, misinformation, and distrust. Effective science communication, employing a "broaden and build approach," can stimulate interest, curiosity, and awe, encouraging additional learning and understanding. Ultimately, presenting information simply and clearly enhances comprehension, trust, retention, application, and interest.

5.6 Recommendations for Effective Communications

Based on the literature reviewed, we provide a series of general recommendations for developing behaviorally informed campaigns to promote climate action. Due to the lack of evidence related to behavioral determinants of climate action in the Argentine population, these recommendations are not specific to Argentina. Still, they are inspired by relevant literature in the field.

To target individual determinants, we suggest the following two recommendations:

1. **Develop Simple Clear Messages**. This includes the use of clear, repeated messages from trusted sources, as well as the promotion of easy, fun, and popular behaviors (Maibach et al. 2023). To improve understanding and decision-making, it is important to reduce the complexity of climate science communication and ensure that key information is conveyed (Amelung et al. 2016). For example, in Southeast England a group of researchers investigated the effectiveness of using stickers as a visual prompt to encourage the separate collection of household food waste for recycling and found sticker prompts to have a significant and sustained positive impact on recycling rates (Shearer et al. 2017).
2. **Use Messages That Give Explicit Moral Explanations**. Conservative audiences are more receptive to messages that refer to authority and civic duty while liberal audiences are more receptive to messages that refer to prosociality and environmental protection (Newell and Moss 2021). Kidwell et al. (2013) developed tailored persuasive messages that appealed to liberals, based on fairness and avoiding harm to others, and conservatives, based on duty and an obligation to

adhere to authority. These congruent appeals significantly affected consumers' acquisition, usage, and recycling intentions and behaviors.

In terms of social factors, we suggest the following:

3. **Draw from Culturally and Locally Relevant Experiences to Create Storytelling That Appeals to the Audience's Emotions**. In line with segmenting the population, understanding the experiences of our audience is key to personalizing and delivering personally relevant messages. For example, after exploring consumer preferences to identify appealing alternative names for vegetarian dishes, Vennard et al. (2019) increased meat-eaters propensity to choose vegetarian food by changing how plant-based dishes are described on menus.
4. **Segment Audiences**. This is a crucial strategy in science communication planning, allowing communicators to understand groups of people with shared knowledge, emotions, values, and behaviors. By categorizing individuals into "motivationally coherent" audience groups, their needs and concerns can be better addressed (Maibach et al. 2023). An example of this approach is the Global Warming's Six Americas segmentation analysis, which classifies the US adult population into six distinct audiences (Maibach et al. 2011; Chryst et al. 2018). This method has been utilized in climate communication initiatives, and a concise survey tool is available for identifying the prevalence of the Six Americas in various populations (Leiserowitz et al. 2021; Roser-Renouf et al. 2022).

Finally, to address physical barriers, we propose the following two measures:

5. **Use Policies in Which the Default Option is the Desired Option**. Reduce the friction for those who are choosing sustainable options. By facilitating the desired action, we leverage the reduced cognitive bandwidth of people who may not be interested in making a conscious decision on what choice to make. For example, a randomized controlled trial in Germany tested the impact of default rules on voluntary purchases of "green" energy contracts. Setting the default choice to more expensive "green" energy increased its purchases nearly tenfold (Ebeling and Lotz 2015).
6. **Extrinsic Policy Incentives Should Be Provided Along with Intrinsic Appeals**, making sure the hedonic, gain and normative goals are addressed by sustainable options in tandem. Because climate change is a long-term global environmental problem, viable adaptation and mitigation solutions will require leveraging stable long-term drivers.

5.7 Final Remarks

Behavioral science can play a crucial role in helping people engage with climate action. By understanding human behavior, policymakers can—and should— tailor their approaches to encourage more sustainable behaviors and combine these insights with practical knowledge across different dimensions. The most straightforward implementation of behavioral science is the development of "nudges" to promote sustainable choices, for example, by making climate-friendly behaviors the default option. Other alternatives include minimizing "sludge"—reducing frictions and barriers to climate action, for example, by making sustainable options more available or affordable. However, making it easier to take environmental actions is not enough, since contextual changes can easily bring back the behavioral issue. As a complement, how science is communicated can have a long-lasting impact on people's willingness to behave pro-environmentally.

This book chapter focused on relevant literature that sheds light on designing effective messages to promote climate-friendly behaviors. We first presented the behavioral issue and a framework that allows us to categorize the different determinants of climate action. Then provided evidence on how polarized societies might affect behavioral plasticity. We introduced the existing gap between climate acceptance and climate action and provided an outline of the most relevant determinants of pro-environmental behaviors. Finally, we provided some recommendations that can help decision-makers in crafting effective messages informed by behavioral science.

How information is framed can significantly impact how it is received and acted upon but not all framings appeal to the same populations and are effective for the same behaviors. In order to know what message works best, policymakers should consider the key determinants of the behaviors they are trying to change and design tailored interventions to tackle those determinants effectively. Another contribution behavioral science can make to address the climate crisis is to help control the spread of false information. Research in climate communication and behavior change has highlighted the need for designing effective messages to combat misinformation and disinformation (Ceyhan and Saribas 2022).

In Argentina, we need more and better behavioral science on climate behavior in Argentina. Political polarization endangers the massive support for climate-related policies. Understanding how to address the different groups across the whole political spectrum is key to crafting messages that resonate with their audience and motivate action. While the country seems to be committed to climate change mitigation and adaptation goals, behaviorally informed policies are still scant. However, in the very few cases where interventions were implemented, they showed positive results, demonstrating how behavioral science can help elicit climate action. In summary, behavioral science undoubtedly provides a new opportunity to bring people back to the center of public policy, addressing their needs while promoting a carbon–neutral environment.

References

Abrahamse W (2019) Encouraging sustainable transport choices. In: Encouraging pro-environmental behavior, pp 97–109. https://doi.org/10.1016/b978-0-12-811359-2.00007-x

Black JS, Stern PC, Elworth JT (1985) Personal and contextual influences on household energy adaptations. J Appl Psychol 70:13–21. https://doi.org/10.1037/0021-9010.70.1.3

Cialdini RB (2001) Influence: science and practice, 3rd ed. Harper Collins
Amelung D, Fischer H, Kruse L, Sauerborn R (2016) Defogging climate change communication: how cognitive research can promote effective climate communication. Front Psychol 7:2014–2017. https://doi.org/10.3389/fpsyg.2016.01340
Austin J, Hatfield DB, Grindle AC, Bailey JS (1993) Increasing recycling in office environments: the effects of specific, informative cues. J Appl Behav Anal 26:247–253. https://doi.org/10.1901/jaba.1993.26-247
Bergquist M, Thiel M, Goldberg MH, van der Linden S (2017) Field interventions for climate change mitigation behaviors: a second-order meta-analysis. Proc Natl Acad Sci. https://doi.org/10.1073/pnas
Birch S (2020) Political polarization and environmental attitudes: a cross-national analysis. Env Polit 29:697–718. https://doi.org/10.1080/09644016.2019.1673997
Buehler R, Pucher J, Kunert U (2009) Making transportation sustainable: insights from Germany
Capstick S, Whitmarsh L, Poortinga W et al (2015) International trends in public perceptions of climate change over the past quarter century. Wiley Interdiscip Rev Clim Chang 6:35–61. https://doi.org/10.1002/wcc.321
Ceyhan GD, Saribas D (2022) Research trends on climate communication in the post-truth era. Educ Dev Psychol 39:5–16
Chryst B, Marlon J, van der Linden S et al (2018) Global Warming's "Six Americas Short Survey": Audience Segmentation of Climate Change Views Using a Four Question Instrument. Environ Commun 12:1109–1122. https://doi.org/10.1080/17524032.2018.1508047
Cole JC, Gillis AJ, van der Linden S et al (2023) Social psychological perspectives on political polarization: insights and implications for climate change. Perspect Psychol Sci. https://doi.org/10.1177/17456916231186409
Dietz T, Gardner GT, Gilligan J et al (2009) Household actions can provide a behavioral wedge to rapidly reduce US carbon emissions. Proc Natl Acad Sci USA 106:18452–18456. https://doi.org/10.1073/pnas.0908738106
Ebeling F, Lotz S (2015) Domestic uptake of green energy promoted by opt-out tariffs. Nat Clim Chang 5:868–871. https://doi.org/10.1038/nclimate2681
Espina Mairal S, Bustos F, Solovey G, Navajas J (2024) Interactive crowdsourcing to fact-check politicians. J Exp Psychol Appl 30(1):3
Environics International Environmental Monitor, (2000) Environics analytics, Toronto. https://doi.org/10.5683/SP3/IAVDO4. Accessed 30 May 2013
Falkenberg M, Galeazzi A, Torricelli M et al (2022) Growing polarization around climate change on social media. Nat Clim Chang 12:1114–1121. https://doi.org/10.1038/s41558-022-01527-x
Feldman L, Hart PS (2018) Climate change as a polarizing cue: framing effects on public support for low-carbon energy policies. Glob Environ Chang 51:54–66. https://doi.org/10.1016/j.gloenvcha.2018.05.004
Fengler W (2022) Could 2022 be a year of hope for climate?
Fisher DR, Waggle J, Leifeld P (2013) Where does political polarization come from? Locating polarization within the U.S. climate change debate. Am Behav Sci 57:70–92. https://doi.org/10.1177/0002764212463360
Flores A, Cole JC, Dickert S et al (2022) Politicians polarize and experts depolarize public support for COVID-19 management policies across countries. Proc Natl Acad Sci USA. https://doi.org/10.1073/pnas.2117543119
Freira L, Sartorio M, Boruchowicz C et al (2021) The interplay between partisanship, forecasted COVID-19 deaths, and support for preventive policies. Humanit Soc Sci Commun 8:1–10. https://doi.org/10.1057/s41599-021-00870-2
Fumagalli E, Krick CB, Dolmatzian MB et al (2023) Partisanship predicts COVID-19 vaccine brand preference: the case of Argentina. Humanit Soc Sci Commun. https://doi.org/10.1057/s41599-023-02067-1

Guagnano GA, Stern PC, Dietz T (1995) Influences on attitude-behavior relationships: a natural experiment with curbside recycling. Environ Behav 27:699–718. https://doi.org/10.1177/0013916595275005

Global Carbon Budget (2023) Population based on various sources (2023). Extracted from: https://ourworldindata.org/co2/country/argentina

GlobeScan (2006) Trust in the media. Available via BBC/Reuters/Media Center Poll. www.globescan.com/news_archives/Trust_in_Media.pdf. Accessed 22 April 2013

Griffith-Jones S, Tyson J (2013) The European Investment Bank: lessons for developing countries (No. 2013/019). WIDER Working Paper

Hart PS, Nisbet EC (2012) Boomerang effects in science communication: how motivated reasoning and identity cues amplify opinion polarization about climate mitigation policies. Communic Res 39:701–723. https://doi.org/10.1177/0093650211416646

Heller K, Berger M, Gagern A et al (2021) Six Behaviors Policymakers Should Promote to Mitigate Climate Change. Behav Sci Policy 7:63–73. https://doi.org/10.1177/237946152100700206

He F, Duan L, Cao Y, Wen S (2024) Green credit policy and corporate climate risk exposure. Energy Econ 133:107509

Hotspot HC (2021) Waste management in the LATAM region. Waste Management Country Report, Mexico

IEA Data Services (May, 17, 2022). Extracted from: https://www.iea.org/countries/argentina/efficiency-demand

Intergovernmental Panel on Climate Change (2018) Global warming of 1.5 °C. In: Masson-Delmotte V, Zhai P, Pörtner H-O, Roberts D, Skea J, Shukla AP, Moufouma-Okia W, Péan C, Pidcock R, Connors S, Matthews JBR, Chen Y, Zhou X, Gomis MI, Lonnoy E, Maycock MT (eds) Special report on global warming of 1.5 °C. Cambridge University Press, Cambridge, UK and New York, NY, USA, pp 228–231

IPCC (2022) IPCC 2022: Summary for policymakers. In: Shukla PR, Skea J, Reisinger A et al (eds) Climate change 2022: mitigation of climate change. Contribution of working group III to the sixth assessment report of the intergovernmental panel on climate changes. Cambridge University Press, Cambridge, UK and New York, NY, USA, pp 1–52

Jakovcevic A, Steg L, Mazzeo N, Caballero R, Franco P, Putrino N, Favara J (2014) Charges for plastic bags: motivational and behavioral effects. J Environ Psychol 40:372–380

Keith Chen M, Rohla R (2018) The effect of partisanship and political advertising on close family ties. Science (80-) 360:1020–1024. https://doi.org/10.1126/science.aaq1433

Kidwell B, Farmer A, Hardesty DM (2013) Getting liberals and conservatives to go green: political ideology and congruent appeals. J Consum Res 40:350–367. https://doi.org/10.1086/670610

Kukowski CA., Hofmann W, Roozenbeek J, van der Linden S, Vandenbergh MP, Nielsen KS (2023) The perceived feasibility of behavior change is positively associated with support for domain-matched climate policies. One Earth 6(11):1554–1563

LeDoux JE (1989) Cognitive—emotional interactions in the brain. Cogn Emot 3:267–289. https://doi.org/10.1080/02699938908412709

Leiserowitz A, Roser-Renouf C, Marlon J, Maibach E (2021) Global Warming's Six Americas: a review and recommendations for climate change communication. Curr Opin Behav Sci 42:97–103. https://doi.org/10.1016/j.cobeha.2021.04.007

Maibach EW, Leiserowitz A, Roser-Renouf C, Mertz CK (2011) Identifying like-minded audiences for global warming public engagement campaigns: an audience segmentation analysis and tool development. PLoS ONE. https://doi.org/10.1371/journal.pone.0017571

Maibach EW, Uppalapati SS, Orr M, Thaker J (2023) Harnessing the power of communication and behavior science to enhance society's response to climate change. Annu Rev Earth Planet Sci 51:53–77. https://doi.org/10.1146/annurev-earth-031621-114417

Marx SM, Weber EU, Orlove BS et al (2007) Communication and mental processes: experiential and analytic processing of uncertain climate information. Glob Environ Chang 17:47–58. https://doi.org/10.1016/j.gloenvcha.2006.10.004

McConnell C, Margalit Y, Malhotra N, Levendusky M (2018) The economic consequences of Partisanship in a polarized era. Am J Pol Sci 62:5–18. https://doi.org/10.1111/ajps.12330
McEvoy D, Fünfgeld H, Bosomworth K (2013) Resilience and climate change adaptation: the importance of framing. Plan Pract Res 28:280–293. https://doi.org/10.1080/02697459.2013.787710
Mertens S, Herberz M, Hahnel UJJ, Brosch T (2022) The effectiveness of nudging: a meta-analysis of choice architecture interventions across behavioral domains. Proc Natl Acad Sci USA 119:1–10. https://doi.org/10.1073/pnas.2107346118
Mols F, Haslam SA, Jetten J, Steffens NK (2015) Why a nudge is not enough: A social identity critique of governance by stealth. Eur J Polit Res 54:81–98. https://doi.org/10.1111/1475-6765.12073
ND-GAIN Country Index (2023) The Notre Dame global adaptation initiative. https://gain.nd.edu/our-work/country-index/rankings/. Accessed 27 Feb 2024
Newell BR, Moss J (2021) Making it easier to take environmental actions is not enough: Policymakers must also emphasize why action is necessary. Behav Sci Policy 7:91–99. https://doi.org/10.1353/bsp.2021.0015
Nielsen KS, Cologna V, Bauer JM, Berger S, Brick C, Dietz T, Wolske KS et al (2024) Realizing the full potential of behavioural science for climate change mitigation. Nat Clim Change:1–9
Nisa CF, Bélanger JJ, Schumpe BM, Faller DG (2019) Meta-analysis of randomised controlled trials testing behavioural interventions to promote household action on climate change. Nat Commun. https://doi.org/10.1038/s41467-019-12457-2
Pegels A, Castañeda JL, Humphreys C et al (2022) Aligning recycling behaviors and the recycling system—towards a full cycle of materials and behavioral methods. Waste Manag 138:1–7. https://doi.org/10.1016/j.wasman.2021.11.021
Poushter J, Huang C (2019) Climate change still seen as the top global threat, but cyberattacks a rising concern
Ramos Muñoz D (2023) Integrating climate risk in banking regulation. University of Genoa EUSFiL Law Research Working Paper Series (20)
Rode JB, Dent AL, Benedict CN et al (2021) Influencing climate change attitudes in the United States: a systematic review and meta-analysis. J Environ Psychol 76:101623. https://doi.org/10.1016/j.jenvp.2021.101623
Roser-Renouf C, Myers TA, Luong KT, Maibach EW (2022) Messages targeted to global warming's six Americas can increase their issue engagement. Res Sq
Shearer L, Gatersleben B, Morse S et al (2017) A problem unstuck? Evaluating the effectiveness of sticker prompts for encouraging household food waste recycling behaviour. Waste Manag 60:164–172. https://doi.org/10.1016/j.wasman.2016.09.036
Steg L (2023) Psychology of climate change. Annu Rev Psychol 74:391–421. https://doi.org/10.1146/annurev-psych-032720-042905
Steg L, Bolderdijk JW, Keizer K, Perlaviciute G (2014) An integrated framework for encouraging pro-environmental behaviour: the role of values, situational factors and goals. J Environ Psychol 38:104–115. https://doi.org/10.1016/j.jenvp.2014.01.002
Stern PC (2016) Sociology: impacts on climate change views. Nat Clim Chang 6:341–342. https://doi.org/10.1038/nclimate2970
Sunstein CR (2016a) Do people like nudges? Adm Law Rev 68:177
Sunstein CR (2016b) The ethics of Influence, 1st edn. Cambridge University Press, Massachusetts
Van Bavel JJ, Pereira A (2018) The Partisan brain: an identity-based model of political belief. Trends Cogn Sci 22:213–224. https://doi.org/10.1016/j.tics.2018.01.004
van der Linden S, Maibach E, Leiserowitz A (2015) Improving public engagement with climate change: five "best practice" insights from psychological science. Perspect Psychol Sci 10:758–763. https://doi.org/10.1177/1745691615598516
van Valkengoed AM, Abrahamse W, Steg L (2022) To select effective interventions for pro-environmental behaviour change, we need to consider determinants of behaviour. Nat Hum Behav 6:1482–1492. https://doi.org/10.1038/s41562-022-01473-w

Vennard D, Park T, Attwood S (2019) Encouraging sustainable food consumption by using more appetizing language. Washington, DC

Vlasceanu M, Coman A (2018) Mnemonic accessibility affects statement believability: the effect of listening to others selectively practicing beliefs. Cognition 180:238–245. https://doi.org/10.1016/j.cognition.2018.07.015

Vlasceanu M, Doell KC, Bak-Coleman JB et al (2024) Addressing climate change with behavioral science: a global intervention tournament in 63 countries. Sci Adv 10:eadj5778. https://doi.org/10.1126/sciadv.adj5778

Whitmarsh L (2008) Are flood victims more concerned about climate change than other people? The role of direct experience in risk perception and behavioural response. J Risk Res 11:351–374. https://doi.org/10.1080/13669870701552235

Wolske KS, Stern PC (2018) Contributions of psychology to limiting climate change. In: Psychology and climate change: human perceptions, impacts, and responses. Elsevier Inc., pp 127–160

World Bank Group (2022) Country climate and development report

World Resources Institute (2022) Climate watch (CAIT)—country greenhouse gas emissions data. https://datasets.wri.org/dataset/cait-country

Chapter 6
Energy Efficiency and Climate Adaptation: A Behavioral Lens on Informal Settlements Housing in Argentina

Gastón Gertner and Cinthia Goytia

Abstract Argentina's pursuit of climate resilience requires accelerating its transition to efficient energy management—especially within under-resourced urban areas. This chapter examines how a behavioral approach can shed light on the complexities of energy efficiency in informal settlements in Buenos Aires (CABA), where poverty, housing quality, and cultural norms intersect. By investigating the practices and architectural choices that shape energy use, our study highlights the critical role that behavioral insights can play in enhancing both living conditions and climate adaptation. We integrate these findings into a holistic framework—one that fuses urban planning with behavioral economics—to inform policymakers on strategies for reducing emissions and increasing energy savings. To contextualize Argentina's challenges, we draw on international case studies, offering a comparative lens that underscores the broader implications for Latin America. Ultimately, this chapter seeks to narrow the literature gap and guide future policies, ensuring that energy efficiency and climate adaptation work hand in hand to promote inclusive development.

Keywords Housing · Energy efficiency · Argentina · Latin America · Behavioral economics · Climate change

G. Gertner (✉)
School of Transnational Governance, European University Institute, Florence, Italy
e-mail: gaston.gertner@eui.eu

C. Goytia
Centro de Investigación de Política Urbana y Vivienda, Escuela de Gobierno, Universidad Torcuato Di Tella, Buenos Aires, Argentina
e-mail: cgoytia@utdt.edu

E. Belfiori and G. Gertner (eds.), *Climate Change and Economic Growth*,
SpringerBriefs in Latin American Studies, https://doi.org/10.1007/978-3-031-99637-5_6

6.1 Introduction

In the heart of Latin America's informal settlements, a critical challenge unfolds daily: millions of families struggle not just for shelter, but for access to reliable, affordable, and sustainable infrastructure services, such as energy. This challenge represents both an urgent humanitarian concern and a vital opportunity in the global fight against climate change. As cities grapple with the intensifying impacts of climate change, informal settlements—home to over 30% of the urban population in Latin America's largest cities—stand at the frontlines of energy poverty and climate vulnerability.

The scale of this challenge is staggering. Globally, 1.6 billion people live in inadequate housing, with 1.1 billion residing in informal settlements, representing nearly a quarter of the world's urban population (UN Habitat 2024). These communities face a double burden: they are both the most vulnerable to climate change impacts and the least equipped to adapt, particularly in terms of energy infrastructure and efficiency. Cities account for 70–80% of global greenhouse gas emissions while consuming more than three-quarters of global energy (IEA 2024), making urban energy efficiency a crucial lever for climate action.

According to a study led by C40 & Arup (2016), potential actions in cities could achieve up to 40% of the emissions reductions necessary to limit global warming to 1.5 °C, suggesting an important role for cities in climate change mitigation. Meanwhile, developing countries suffer the most as they face heavier barriers on economic growth due to natural disasters, compared to developed countries (Gonzalez 2021). This susceptibility is even worse for informal settlements since they have the least appropriate energy and the most poorly constructed buildings.

Although most policies emphasize the impact of climate change and energy efficiency in formal urban areas, it is vital to acknowledge that informal settlements are also deeply impacted by climate change and face significant challenges in accessing and using energy efficiently. This issue is especially pertinent in Latin America, where the combined pressures of rapid urbanization and climate change are most acute.

The Metripolitan Area of Buenos Aires exemplifies both the challenges and opportunities in addressing urban energy efficiency. As a metropolis that contributes 37% of Argentina's GDP, the city's energy landscape reflects stark disparities between formal and informal neighborhoods. While affluent districts enjoy reliable energy infrastructure, informal settlements—where a significant portion of the population resides—struggle with precarious connections, inefficient housing structures, and unsustainable energy consumption patterns. The residential sector alone accounts for 27% of Argentina's total greenhouse gas emissions (Annual Statistical Report of the Secretariat of Energy of the Argentine Republic 2022), with informal settlements bearing disproportionate energy burdens due to poorly constructed housing and inefficient appliances.

Recent progress in renewable energy adoption across Latin America offers hope. According to the Inter-American Development Bank (IDB), between 2015 and 2022, the region increased its renewable energy capacity by 51%, reaching 64% of generation from renewable sources in 2022. However, this transition needs to be much swifter in order to sustain rising energy requirements and make carbon emissions neutral by the year 2050. These are also the areas where energy issues must be addressed in order to make cities sustainable and more resilient, which is why integrated solutions that link energy, urban design, and community are important.

The challenge is particularly acute in informal settlements, where the complex interplay of social, economic, and technical factors requires innovative, integrated solutions that link energy efficiency, urban design, and community engagement. About 16.8% of the urban population in Latin America lives in informal settlements (UN Habitat 2024), but at least 30% of the population of Latin American largest cities live in informal settlements (ECLAC 2020). These figures are indicative of extreme inequalities in the distribution of social and economic resources across the region (Economic Commission for Latin America and the Caribbean, ECLAC 2020). Most of these settlements are characterized by substandard construction materials that fail to provide adequate thermal insulation and are often deprived of essential resources like clean water, energy, and sanitation; this makes their inhabitants depend on unsustainable energy types like fossil fuels and biomass for burning, which exacerbates greenhouse emission and pollution (Redeus 2021). The urban poor, particularly residents of informal settlements, are disproportionately exposed to extreme weather events on account of their location, poor quality of construction, and limited savings (IPCC 2022). The limited provision of essential services and infrastructure also increases the exposure of these communities to impact from climate change such as landslides and floods (UN Environment 2019a, b; UNDRR 2019; IPCC 2019; Hoegh-Guldberg et al. 2018.). Attention to the provision of basic needs for informal settlements is critical in terms of minimizing climate emissions in the region and enhancing its climate change adaptation strategies. For this to happen, there is need to adopt public policies aimed at regularizing land, integrating them to the cities (Goytia and Dorna 2019), improving physical infrastructure (Goytia et al. 2022), regularizing land (Galiani and Schardgrosky 2012), and making available clean and affordable energy to ensure more equitable and sustainable urban development (Goytia and Gertner 2025; Smolka and Larangeira 2008).

This chapter examines the intersection of energy efficiency, climate adaptation, and behavioral economics in informal settlements, with a particular focus on Argentina's urban housing sector. In Argentina, the residential sector accounts for 27% of total greenhouse gas (GHG) emissions. Household energy consumption is primarily divided among three sources: natural gas, electricity, and, to a lesser extent, liquefied petroleum gas (LPG). Natural gas predominates in urban areas, where it is used for heating, hot water, and cooking. Space heating drives 48% of household energy consumption, followed by water heating (21%) and cooking (15%). Appliances like refrigerators and freezers represent a significant share of electricity consumption, particularly in low-income households. Geographic climate differences influence energy consumption patterns, with higher per capita energy use in the colder regions of the country. Human quality of life depends on thermal comfort: the use of stoves, hot showers, and food preservation. Each of these activities contributes to CO_2 emissions. Energy inefficiency is a serious issue, with many households relying on outdated appliances that consume far more energy than modern, efficient models. The cost of energy affects households unequally, depending on income levels and climate variability. Segmenting consumers by income levels (low, middle, and high) is a best-practice strategy to adjust electricity tariffs and subsidies, targeting support for the most vulnerable sectors while reducing subsidies for higher-income consumers.

As the City of Buenos Aires and its Metropolitan Area expands, it faces the pressing need to address climate change and energy efficiency challenges. Affluent districts have reliable energy infrastructure, while low-income areas, especially informal settlements, face limited access to all basic services (Goytia et al. 2022). Poorly built homes with inefficient appliances rely on unsafe, informal energy sources for which the families are not billed individually, leading to higher consumption, unaffordable formal connections, and a cycle of inefficiency and high costs (Goytia et al. 2023b). So, what is the suggested policy pathway for Argentina, and other countries in the region, to achieve a sustainable and equitable future within its formal vs. informal social and spatial divide across cities?

Integration climate action into urban planning and building construction design frameworks is essential for a sustainable future. Advances in behavioral economics and behavioral offer valuables approaches to better understand the context in which they apply decision-making for energy consumption patterns. Insights from behavioral economics can also assist in the development of strategies that overcome hurdles to the uptake of energy conserving technologies and practices such as energy-efficient cook stoves and improved insulation materials. For instance, behavioral nudge strategies, in the form of voluntary commitments, change of default options or campaigns encouraging the use of sustainable technology can lead to energy savings resulting from energy-efficient and conservation measures. By applying these behavioral strategies in combination with urban and architectural design, real transformation processes shall become possible in the targeted communities for the long time.

This chapter aims to explore the different forms in which behavioral science ideas and concepts are connected with current fundamental priorities, policy interventions, and initiatives within urban planning, housing solutions, and nudge-based interventions to propel adaptation and mitigation policies to face climate change challenges in cities and the residential sector in a context of energy transition.

Through a behavioral lens, we explore how choice architecture, which means the deliberate crafting of decision-making environments by subtly shaping how options are presented, when adding to behavioral insights, can inform more effective interventions in informal settlement housing. Our analysis draws on detailed case studies from Buenos Aires' informal settlements and new social housing projects, where striking contrasts in energy consumption patterns reveal both challenges and opportunities for improvement.

The chapter is structured as follows: first, we examine residential greenhouse gas emissions and urban planning solutions through a behavioral economics framework. We then explore how choice architecture principles can be applied to urban planning and housing design to promote energy efficiency. The following sections analyze the role of default energy-saving mechanisms in housing construction, the importance of energy literacy and environmental awareness, and the critical function of energy-efficient appliances. Throughout, we integrate findings from innovative pilot programs and field studies that demonstrate the potential for transformative change when behavioral insights are combined with technical solutions. We conclude with actionable recommendations for policymakers, urban planners, and community stakeholders, emphasizing the importance of integrated approaches that address both the physical and behavioral dimensions of energy efficiency in informal settlements.

By bringing together behavioral science, urban planning, and energy efficiency, this chapter offers a novel framework for understanding and addressing the complex challenges of sustainable development in informal settlements. Our analysis suggests that successful interventions must combine technical solutions with deep understanding of human behavior, community dynamics, and local context to achieve lasting improvements in energy efficiency and climate resilience.

6.2 Understanding Residential GHG Emissions, Urban Planning, and Policy Solutions Through a Behavioral Lens

In recent years, the integration of cognitive sciences into design, architecture, and urban planning for shaping built environments and optimize spatial design has grown significantly, giving rise to fields such as behavioral architecture and urbanism. These disciplines explore how human behavior can be influenced through design and policy, emphasizing non-coercive methods like nudging—creating conditions that guide individuals toward desired actions without imposing them. Policymakers and academics alike are increasingly incorporating behavioral insights to design public services that encourage collective actions and sustainable practices.

Behavioral approaches to urban planning and policy can be framed through two complementary perspectives: behavioral urbanism (BU), as conceptualized by Heimsath (1977), examines the interactions between people and the built environment, focusing on how urban design impacts cognitive, emotional, and social behaviors. It explores the behavioral consequences of urban structures designed with traditional engineering and aesthetic criteria, often without consideration for how these choices influence well-being or encourage behavioral change.

In second place, Behavioral economics is grounded in the work of Thaler and Sunstein (2008) and further developed by Oliver (2017) and Viale (2018, 2022), as it applies cognitive science and fosters the design of behaviorally informed interventions through nudges to affect public policy impact. Behavioral economics aims to motivate behavioral changes that promote individual and collective well-being across areas such as education, health, mobility, energy use, and environmental sustainability. However, it has largely overlooked urban planning, focusing instead on public services with spatial implications for city design.

A notable gap exists between these two approaches. Behavioral urbanism studies the impact of pre-determined urban plans on behavior, but it seldom evaluates how to use behavioral insights to shape these plans proactively. Meanwhile, behavioral economics rarely addresses the spatial dimensions of cities, leaving urban planning disconnected from its behavioral goals. Bridging this divide is essential for developing integrated strategies that combine urban and architectural design with behavioral science to enhance human well-being. An increase in behaviorally informed frameworks can offer a more holistic approach to addressing challenges such as residential GHG emissions, fostering sustainable urban environments that align policy solutions with behavioral dynamics.

6.3 Choice Architecture in Urban Planning

A key tool for creating sustainable cities is urban planning, especially when it comes to tactics that affect land use, mobility, and energy use. From the field of behavioral economics, "choice architecture" has become popular as a way to direct urban decision-making and encourage environmentally friendly practices. Choice architecture can encompass how urban planners create neighborhood environments; for example, designing walkable communities dotted with strategically located facilities to encourage sustainable travel through improved public transit and bike-friendly infrastructure, as well as lowering barriers to adopting environmentally conscious behaviors. In keeping with this idea, the placing of schools, markets, and healthcare facilities within walking distance not only reduces car dependency but also greenhouse gas emissions. Beauty and usability in pedestrian pathways and bike lanes make them automatic options in the travel system, which helps to prevent reliance on automobiles while encouraging mobility modes substitution effects.

In exploring choice architecture usage by urban planners in reforming ambient neighborhoods, Mehaffy (2017) focuses on strategies like walkable communities with their amenities located conveniently, advocating for sustainable mobility with improved public transit and bike lanes infrastructure, and lowering the threshold to adopting eco-friendly practices. For example, the locations of schools, markets, and health centers place themselves within walking distances which reduce dependency on cars and, therefore, greenhouse gas emissions. In Tokio, these services are located at less than 500 mts from public transportation stations, to induce greater use of mass mobility modes, which reduced emissions, for all activities. Similarly, redesigning and making pedestrian pathways and bike lanes functional makes them default modes of transport by which humans are prevented from becoming car dependent.

This approach is driven by behavioral insights, for example, framing incentives in such a way that they communicate immediate and safe benefits, addressing cultural perceptions, and making it easy and visible to people for making sustainable choices. Kunreuther and Weber (2013) show that indeed much attention should be paid to defaults and framing as an avenue for overcoming cognitive barriers toward embracing long-term sustainable practices. However, cities that combine urban design principles with behavioral economic tools will not only mitigate adverse environmental impacts but also improve the quality of life for its residents affords direct savings toward more resilient futures.

In urban planning, choice architecture offers a potent instrument for promoting sustainable urban growth. Creating neighborhoods with well-located resources, such markets, schools, and medical facilities, within walking distance is a crucial tactic. This promotes healthier, community-focused lifestyles, lessens reliance on cars, and lowers greenhouse gas emissions. According to Mehaffy (2017), short blocks and a high intersection density improve the usability of urban areas by increasing the accessibility and appeal of walking and bicycling.

Investing in sustainable mobility infrastructure is another important factor. Effective public transportation networks, bike lanes, and pedestrian walkways promote the transition away from driving a car and toward more environmentally friendly forms of transportation. Mehaffy (2017) emphasizes the significance of appealing and practical designs, such as aesthetically pleasing routes and conspicuous transit stops, which inherently become the default mode of transportation and further lower automobile emissions by driving mode substitution effects from automobile to mass public transportation.

By tackling the mental and cultural obstacles to implementing sustainable practices, behavioral economics concepts enhance these architectural plans. It is easier to promote long-term behavior changes when incentives are framed to highlight immediate advantages like ease and cost savings. Kunreuther and Weber (2013) emphasize the importance of nudges and defaults in encouraging people to make environmentally responsible decisions. When combined, these strategies show how incorporating choice architecture into urban design can result in more resilient and habitable communities that are also environmentally friendly.

6.4 Solving the Energy Divide in Buenos Aires

6.4.1 Energy Saving by Default: Building Energy-Efficient Housing

The right to adequate housing is a cornerstone of Argentina's legal framework, sealed in the National Constitution and imprinted in international treaties by which Argentina abides. At a subnational level, the constitution of the City of Buenos Aires extends this right beyond shelter to include permanent access to natural resources through public services such as potable water and reliable energy for cooking. Energy use in residential buildings accounts for approximately 28% of Buenos Aires' total greenhouse gas (GHG) emissions,[1] making the sector a critical target for energy efficiency and sustainable policies. Since the launch of the City of Buenos Aires Climate

[1] Greenhouse Gas Inventory (GHGI) 2018.

Action Plan 2050, a key element of the strategy has been incorporating energy efficiency measures into the improvement of informal settlements and the construction of new social housing.[2] This approach acknowledges that achieving carbon neutrality requires innovative solutions that not only lower emissions but also improve the quality of life for all residents, especially those in vulnerable communities.

A major challenge in informal settlements is the substandard construction materials that fail to provide adequate thermal insulation, leaving interiors uncomfortably hot in summer and cold in winter. Many homes are built by informal settlers with inadequate insulation, leaky roofs, and poor ventilation, resulting in structures ill-equipped to withstand harsh weather conditions. Consequently, households rely on inefficient and expensive means of heating and cooling, straining both their budgets and the energy grid. While formalizing energy connections and bringing access to the energy grid for lower-income households is crucial, it comes with challenges. Outdated appliances, bad construction quality of the houses, and limited awareness of energy efficiency can push these households beyond the reach of basic (affordable) energy tariffs. If formality becomes costly, households may resort to informal connections once again, perpetuating a cycle of marginalization and low quality of life.

New social housing programs or the upgrading of existing informal units meets high-efficiency standards while integrating them into the formal grid.

Sustainable materials are essential for mitigating these extremes, enabling homes to maintain stable indoor temperatures, and reducing reliance on energy-intensive heating and cooling systems. Reducing dependency on mechanical climate control systems requires the implementation of passive design techniques, such as the thoughtful placement of windows and the use of translucent materials to optimize natural lighting and airflow. In addition to improving thermal comfort, these methods lower energy use and related pollutants. The use of passive solutions is especially pertinent in informal settlements, where infrastructure is unstable and resources are scarce.

Efficient housing solutions leverage choice architecture by embedding energy-saving behaviors into the very design of building infrastructure, leaving minimal room for inefficient decisions. In informal settlements, where homes often lack insulation and feature poor construction quality, the dependency on mechanical heating and cooling systems drives up energy costs and emissions. By embedding these solutions into housing design, residents inherently benefit from reduced energy consumption without needing to actively make these choices. This "default option" strategy promotes energy efficiency as a baseline standard, aligning with behavioral economics principles to drive sustainable living habits.

Fernández Reyna (2008) explores how passive design can improve the energy efficiency and livability of informal urban structures. The study highlights that even in resource-constrained environments, techniques such as using locally sourced materials and optimizing window orientation can significantly enhance thermal comfort.

[2] Climate Action Plan 2050: City of Buenos Aires (2023), https://buenosaires.gob.ar/sites/default/files/2023-02/pac_2050_buenos_aires_eng_0.pdf.

Similarly, Cuesta Beleño and Buitrago (2020) examine the application of these strategies within the context of territorial planning and neighborhood regularization, emphasizing that passive design not only boosts energy efficiency but also fosters sustainability and better living conditions in underserved communities. Both studies show the value of using passive design techniques in informal areas, emphasizing how they can maximize thermal comfort, lower energy use, and advance sustainability in under-infrastructured neighborhoods.

Under the Approval Act for the Sustainable Housing and Habitat Strategy (EVHS) of Buenos Aires, established by the Housing Institute of the City (IVC), new housing developments in the Rodrigo Bueno and Fraga neighborhoods (Playón de Chacarita) were designed with environmental criteria that enhance energy efficiency and incorporate renewable energy systems to reduce energy consumption for heating water. Projects such as Rodrigo Bueno aim to counter this by implementing passive design techniques, such as strategic window placement and the use of sustainable materials, ensuring thermal comfort by default. A pilot study by Goytia et al (2023a, b) estimated a 10–23% Initial Primary Energy (IPE)[3] demand levels decrease between a sample of new dwellings in Rodrigo Bueno and Fraga neighborhoods financed by this program and mean IPE estimates for aggregate conglomerates in the Cities of Rosario and Buenos Aires. Lower IPE values indicate more energy-efficient and sustainable buildings, as they require less primary energy over their lifecycle.

All in all, beneficiary households in new social housing enjoyed safer, more reliable energy services, while un-upgraded households within the same neighborhoods faced frequent outages, hazardous connections, and the risk of disconnection. The analysis showed that 40–60% of energy use was spent on maintaining indoor comfort. Informal housing not only consumed more energy than new social housing but also had poorer thermal conditions, increasing costs for those connected to formal networks with monthly billing.

Informal neighborhoods residents' health—particularly the poorest and most marginalized—is adversely affected by substandard housing conditions as well as restricted access to essential utilities like electricity, water, and sanitary facilities (Niragire 2023). As a consequence, raising the standard of building materials not only helps to lower greenhouse gas emissions but also improves the quality of life for locals by creating more comfortable, affordable and healthy indoor spaces.

6.4.2 Energy Literacy, Environmental Awareness and Energy-Saving Behaviors

The prevalence of energy-saving behaviors is particularly low in cities where prices for energy use are subsidized. Thus, controlling water use for cooking and cleaning

[3] Initial Primary Energy (IPE) represents the total primary energy needed to operate and maintain a building, accounting for both direct energy use (such as heating and lighting) and indirect energy use (including the energy embedded in construction materials).

and setting a default temperature for air conditioning in the summer becomes a challenge for most households. Particularly, it is easy for energy efficiency initiatives in informal settlements to be lost in the challenging demands of daily life in informal settlements—sourcing water, managing irregular employment opportunities, or basic health and safety concerns, compounded by risk aversion and the discounting of future-based benefits (Figueroa et al. 2019).

Accessing and understanding energy consumption at the household level are key for residents to engage in energy-saving strategies that also reduce GHG. To this end, residents, at a basic level, should be able to read basic consumption levels information detailed in electricity bills. Further, they should be able to distinguish patterns of energy use link to heating, cooling, and other household appliances' use.

In Argentina, energy literacy and awareness levels remain low and are more acute in informal settlements. As part of the pilot study in informal settlements within the City of Buenos Aires (Goytia et al. 2023b), interview analysis with residents revealed key challenges beyond the scope of building conditions. The study uncovered a significant knowledge gap in understanding energy consumption, efficiency measures, and the long-term benefits of energy-saving practices. Many residents were unfamiliar with basic energy conservation principles, such as turning off unused lights and appliances, utilizing natural ventilation instead of air conditioning, or choosing energy-efficient devices. This knowledge deficit was compounded by limited access to resources and information on energy management. Barriers such as language differences, low literacy levels, and the absence of targeted outreach programs left many households without reliable guidance. The findings emphasized the critical need for culturally appropriate and accessible energy education programs that empower communities with practical knowledge and tools to adopt efficient practices.

The study also found entrenched behavioral patterns that drive energy inefficiency. Common practices included leaving appliances on standby mode and excessive daytime lighting with bulbs on. These behaviors, often influenced by cultural norms and social expectations, presented challenges for energy-saving interventions. For example, certain appliances or maintaining specific indoor temperatures were associated with social status or hospitality, even if they led to higher costs. In some cases, keeping lights and televisions on all day served as a substitute for the sense of security lacking in these communities, reflecting broader systemic vulnerabilities.

Overcoming these hidden barriers demands a fundamental shift in the approach to energy efficiency interventions. Participatory, community-driven strategies that actively involve residents in the process of energy transformation can have significant impacts in rising awareness and removing barriers to energy-saving behaviors. This required moving away from top-down, technocratic solutions and instead fostering ownership and empowerment among low-income households. Creating spaces for dialogue, knowledge exchange, and collective action could bridge the gap between technical expertise and local knowledge. By framing energy efficiency as a pathway to broader benefits—such as improved health, greater comfort, and financial stability—interventions could better align with residents' lived experiences, making them more relevant and impactful.

6.4.3 The Role of Energy-Efficient Appliances

Adopting energy-efficient products, such as LED lighting, energy-rated freezers, and modern air conditioners, is a crucial tactic for reducing household energy waste while preserving functionality (Olatunde et al. 2022). These appliances, which are frequently marked with certifications like ENERGY STAR in the US or EU energy labels, are made to use less energy thanks to cutting-edge technologies like variable-speed motors and better insulation. Argentina has implemented specific programs such as the *Energy Efficiency Labeling Program*[4] and the *Renovate Program*,[5] which make it easier to acquire energy-efficient appliances through subsidies and financing.

However, several obstacles prevent energy-efficient appliances from being widely used, especially in cities. Despite the long-term cost benefits, many households are discouraged from investing in these items due to economic issues such as large upfront fees, particularly in low-income neighborhoods. The issue is made worse by low customer knowledge of the advantages of these products and the absence of easily understandable energy efficiency guidelines or labeling schemes. Access to the market is still difficult, especially in informal settlements where it is difficult to find trustworthy vendors of energy-efficient appliances.

Given how behavioral barriers have a big influence on how people make decisions about how much energy to use, they are essential to understanding why energy-efficient habits and technology are adopted slowly. Loss aversion and status quo bias are two cognitive constraints that discourage households from investing in energy-efficient appliances upfront, even when the long-term savings are clear. Because of these prejudices, many are resistant to change and continue to use outdated technology. Short-term thinking also causes many consumers to put immediate cost reductions ahead of the possible long-term advantages of lower energy prices due to efficiency. This is a particular strong challenge in informal settlements where users are not billed for energy use. Hence, their status quo situation is one where they access free energy services at no cost. A lack of easily understandable information combined with the complexity of the available alternatives restricts adoption even more, leaving many customers feeling overloaded.

These difficulties are made worse by financial limitations, particularly for low-income households. Even though energy-efficient appliances provide significant long-term savings, their high initial cost deters investment. Since many consumers are still unaware of the cost and energy advantages of efficient technologies, awareness gaps make the problem worse and need for focused instructional initiatives. Figueroa et al. (2019) offer important insights on addressing these obstacles in their case studies in Nairobi's Kibera. Behavioral nudges like default settings and streamlined decision-making frameworks effectively boost the adoption of energy-efficient technologies in informal settlements. These initiatives address cognitive limits and

---

[4] https://www.argentina.gob.ar/economia/energia/eficiencia-energetica/etiquetado-en-eficiencia-energetica.

[5] https://defensoria.org.ar/archivo_noticias/renovate-se-reglamento-el-programa-de-renovacion-de-electrodomesticos/.

the difficulties presented by flat-rate energy systems, where the financial benefits of lower consumption are less obvious. They are specifically designed to address the conditions of informal settlements. Social norms and pro-social actions also worked well to promote community-driven energy efficiency.

These problems highlight the necessity of encouraging laws, innovations in technology, and educational programs to increase consumer awareness, lower costs, and open up markets.

Energy-efficient appliances not only lower family energy bills but also help to lower carbon emissions and advance environmental sustainability. Addressing these adoption barriers with a unified approach is essential to achieving these benefits. In addition to reducing the financial burden of household energy usage, metropolitan communities can promote a more sustainable, energy-efficient future by enacting explicit energy efficiency standards and labeling, financial incentives, and educational initiatives.

Adoption of energy-efficient technologies has been greatly aided by behavioral interventions, particularly in situations where access is impeded by infrastructure and financial constraints.

The results of the study highlight how crucial it is to incorporate behavioral insights into the creation of policies. While financial incentives were successful in reducing short-term cost barriers, customized nudges that addressed cognitive and informational gaps were necessary for long-term adoption. This illustrates the necessity of a diversified strategy to address the intricate problems that inhabitants in informal communities' encounter.

6.5 Conclusion

The interconnectedness of urban planning, architectural design, and energy consumption patterns underscores the complexity of creating sustainable cities. Urban design shapes energy use through building efficiency and transportation systems, while architectural innovations, such as thermal insulation and passive design, directly influence consumption behaviors. Informal settlements, often overlooked, offer unique opportunities to test behavioral and technological interventions, as their unregulated environments allow for adaptive, context-specific solutions. For public policy, a holistic approach is essential—combining behavioral insights (e.g., nudges), advanced technologies (e.g., smart grids, efficient appliances), and normative program approaches for certified sustainable housing building protocols to drive equitable and impactful change.

In the Latin America and the Caribbean (LAC), which is among the world regions with the most rapid urbanization rates, informal settlements remain an important area for urban growth. All these infrastructures constitute relatively low level of energy efficiency due to poor energy usage practices and malfunctioning infrastructure. Nevertheless, they offer possibilities for sustainable urban development practices. Overcoming unnecessary energy poverty in informal settlements can be achieved through community engagement with regulatory frameworks to create sustainable energy solutions in informal settlements. In the case of LAC, and specifically Argentina, it highlights the importance of place-based and participatory approaches in the pursuit of both urban sustainability and energy equity. However, expanding these initiatives requires further exploration, new policies, and international collaboration for tailoring the approaches to different local settings in the region.

In order to achieve urban sustainability in informal settlements, a multidisciplinary strategy combining behavioral research, energy-efficient technologies, innovative architecture, and urban planning is required. Informal settlements offer special chances to test and scale adaptive, sustainable initiatives, despite the major obstacles presented by uncontrolled building, inadequate infrastructure, and reliance on inefficient energy sources.

Future studies must examine how well these interventions scale in various urban contexts and assess their long-term effects using longitudinal research. Because of their distinct features, informal settlements provide an invaluable environment for testing out novel approaches that promote sustainability. To turn informal settlements into equitable, sustainable urban settings that support more general objectives of resilience and preservation of the environment, cooperation between urban planners, legislators, and communities is essential.

References

Annual Statistical Report of the Secretariat of Energy of the Argentine Republic. Retrieved from https://www.energia.gob.ar/contenidos/archivos/Reorganizacion/planeamiento/publicaciones/iea_2022.pdf

C40 & Arup (2016) Deadline 2020: how cities will get the job done—an analysis of the contribution C40 cities can make to delivering the Paris Agreement objective of limiting global temperature rise to 1.5 degrees. Retrieved from https://www.c40.org/other/deadline_2020

City of Buenos Aires (2023) Climate action plan 2050: city of Buenos Aires. Retrieved from https://buenosaires.gob.ar/sites/default/files/2023-02/pac_2050_buenos_aires_eng_0.pdf

Cuesta Beleño A, Buitrago L (2020) Asentamientos informales como base para la legalización barrial, el diseño y el ordenamiento territorial. Universidad La Gran Colombia. Retrieved from https://repository.ugc.edu.co/bitstream/handle/11396/6050/ASENTAMIENTOS%20INFORMALES.pdf?sequence=1

Economic Commission for Latin America and the Caribbean (ECLAC) (2020) Informal settlements: challenges and public policies in Latin America and the Caribbean. Retrieved from https://repositorio.cepal.org/handle/11362/45632

Fernández Reyna M (2008) Design in urban informal structures Doctoral dissertation, Universitat Politècnica de Catalunya. TDX Repository. Retrieved from https://www.tdx.cat/handle/10803/6813

Figueroa A, de Molière L, Pegels A, Never B, Kutzner F (2019) Show me (more than) the money! Assessing the social and psychological dimensions to energy-efficient lighting in Kenya. Energy Res Soc Sci 47:224–232. https://doi.org/10.1016/j.erss.2018.10.002

Galiani S, Schargrodsky E (2012) Property rights for the poor: effects of land titling. J Public Econ 96(1–2):1–15

Gonzalez F (2021) Desastres naturales y crecimiento económico: una síntesis de la evidencia empírica. Revista de Ciencias Sociales 31(61), enero-junio 2022. https://orcid.org/0000-0002-8347-9808

Goytia S, Dorna M (2019) Land regularization and urban integration policies in Latin America: a comparative analysis. Habitat Int 89:101–110

Goytia C, Gertner (2025) From disparity to sustainability: the journey of energy efficiency in Buenos Aires' low-income neighborhoods. In: Karunaratne G (eds) Resilient urbanism. Routledge research in planning and urban design, forthcoming

Goytia S, Dorna M, Rodríguez M (2022) Improving physical infrastructure in informal settlements: lessons from Latin America. Urban Studies 59(3):567–583

Goytia C, Heikkila EJ, Pasquini RA (2023a) Do land use regulations help give rise to informal settlements? Evidence from Buenos Aires. Land Use Policy 125:106484. https://doi.org/10.1016/j.landusepol.2022.106484

Goytia C, Gertner G, Zavalia R, Gil S (2023b) ESTUDIO DE EFICIENCIA ENERGÉTICA Y DEL AGUA. Programa de Integración Socio-Urbana: Barrio 20, Barrio Rodrigo Bueno y Barrio Fraga de la Ciudad de Buenos Aires. Observatorio de Vivienda de la CABA, Gobierno de la Ciudad de Buenos Aires. Government of the City of Buenos Aires (2021). Greenhouse gas inventory 2018. Retrieved from https://buenosaires.gob.ar/sites/default/files/media/document/2021/06/24/c88b2f9d28277bbb81579e569407ef3c7e8753ba.pdf

Heimsath C (1977) Behavioral architecture: toward an accountable design process. McGraw-Hill, New York

Hoegh-Guldberg O, Jacob D, Taylor M, Bindi M, Brown S, Camilloni I, Diedhiou A, Djalante R, Ebi KL, Engelbrecht F, Guiot J, Hijioka Y, Mehrotra S, Payne A, Seneviratne SI, Thomas A, Warren R, Zhou G (2018) Impacts of 1.5 ºC global warming on natural and human systems. In: Masson-Delmotte V, Zhai P, Pörtner HO, Roberts D, Skea J, Shukla PR, Pirani A, Moufouma-Okia W, Péan C, Pidcock R, Connors S, Matthews JBR, Chen Y, Zhou X, Gomis MI, Lonnoy E, Maycock T, Tignor M, Waterfield T (eds) Global warming of 1.5°C. An IPCC Special Report on the impacts of global warming of 1.5 °C above pre-industrial levels and related global greenhouse gas emission pathways, in the context of strengthening the global response to the threat of climate change, sustainable development, and efforts to eradicate poverty, pp 175–311. World Meteorological Organization. Retrieved from https://www.ipcc.ch/site/assets/uploads/sites/2/2019/02/SR15_Chapter3_Low_Res.pdf

International Energy Agency (2008) World energy outlook 2008. Retrieved from https://iea.blob.core.windows.net/assets/89d1f68c-f4bf-4597-805f-901cfa6ce889/weo2008.pdf

International Energy Agency (2024) Empowering urban energy transitions: smart cities and smart grids. Retrieved from https://iea.blob.core.windows.net/assets/00f7d520-d517-473d-b357-5adb43c4a57e/EmpoweringUrbanEnergyTransitions.pdf

Inter-American Development Bank (2023) Energy transition in Latin America and the Caribbean. Retrieved from https://www.iadb.org/en/news/energy-transition-latin-america-and-caribbean

IPCC (2019) Special report on climate change and land: an IPCC special report on climate change, desertification, land degradation, sustainable land management, food security, and greenhouse gas fluxes in terrestrial ecosystems. Retrieved from https://www.ipcc.ch/srccl/

IPCC (2022) Climate change 2022: mitigation of climate change. In: Working group III contribution to the sixth assessment report of the intergovernmental panel on climate change. Retrieved from https://www.ipcc.ch/report/ar6/wg3/

Kunreuther H, Weber EU (2013) Aiding decision making to reduce the impacts of climate change. Springer

Mehaffy MW (2017) Neighborhood "choice architecture": a new strategy for lower-emissions urban planning? Urban Plann 3(2):113–127

Niragire E (2023) Health and informal settlements: exploring the risks of inadequate housing and limited basic services. Environmental Health News. Retrieved from https://www.ehn.org/health-and-informal-settlements-2659674284.html

Oliver A (2017) The origins of behavioural public policy. Cambridge University Press

Olatunde TM, Okwandu AC, Akande DO (2022) Reviewing the impact of energy-efficient appliances on household consumption. Energy and Policy Research

Redeus (2021) Resilient informal settlements in the face of climate change: Vulnerability and adaptive co-production. Retrieved from https://redeus.org/wp-content/uploads/2021/03/17_Asentamientos-Informales-Resilientes-frente-al-cambio-climatico_-vulnerabilidad-y-coproduccion-adaptativa_Diana-Xie.pdf

Smolka MO, Larangeira A (2008) Regularization of informal settlements in Latin America. Lincoln Institute of Land Policy. Retrieved from https://www.lincolninst.edu/app/uploads/legacy-files/pubfiles/regularizacion-asentamientos-informales-full_0.pdf

Thaler RH, Sunstein CR (2008) Nudge: improving decisions about health, wealth, and happiness. Yale University Press

UN Environment Programme (2019a) Emissions gap report 2019. Retrieved from https://www.unep.org/resources/emissions-gap-report-2019

UN Environment Programme (2019b) UNEP annual report 2019. Retrieved from https://www.unep.org/resources/unep-annual-report-2019

UN-Habitat (2024) World cities report 2024: chapter 4—Housing vulnerabilities and informal settlements. Retrieved from https://unhabitat.org/sites/default/files/2024/11/wcr_2024_-_chapter_4.pdf

UNDRR (2019) UNDRR annual report 2019. Retrieved from https://www.undrr.org/publication/undrr-annual-report-2019

Viale R (2018) Oltre il nudge: libertà di scelta, felicità e comportamento. Società editrice il Mulino

Viale R (2022) Nudging: human behavior and embodied cognition to promote individual and public wellbeing. MIT Press

Chapter 7
Towards a Green Transition: How the Youth is Shaping Economy, Society and Politics

Yanel Llohis, Valentina Sosa Patrignani, and Sofía Berenice Osadczuk

Abstract The twenty-first century has witnessed a rapid increase in interest and engagement concerning environmental degradation and climate change issues, with growing concerns about the harmful consequences these issues may cause in future. This is reflected in a series of events that recently took place in the social, international and economic areas, where the youth have played a central role. This chapter aims to explore the most recent environmental-related shifts in those areas and provide insights into how young people are shaping and leading the transition towards a greener society. It discusses the role of youth in social movements related to environmental protection and the impact of social media in this sphere. Additionally, it delves into recent global political engagement by the youth as a response to the failure of traditional policies. Finally, the chapter analyzes recent shifts in market trends related to environmental products and services and explores how the youth are contributing to this transformation.

Keywords Climate change · Global politics · Green consumption · Social media · Social movements · Youth activism

7.1 Introduction

Young people are certainly leading a change concerning the actions we take concerning the environment. As the consequences of environmental degradation become more evident, with rising temperature levels and worries about abiding by the Paris Agreement, a meaningful change of attitudes aiming to smooth the negative consequences of climate change is taking place worldwide.

Y. Llohis (✉) · S. B. Osadczuk
Universidad Torcuato Di Tella, Buenos Aires, Argentina
e-mail: llohisyanel@gmail.com

V. S. Patrignani
CEPE Di Tella, Buenos Aires, Argentina

E. Belfiori and G. Gertner (eds.), *Climate Change and Economic Growth*, SpringerBriefs in Latin American Studies, https://doi.org/10.1007/978-3-031-99637-5_7

While many individuals today are aware of the importance of protecting the environment and are getting involved in related initiatives, it is the youth who stand as the protagonists. They are driving significant changes in different areas through their decisions and positively influencing others. Moreover, they are actively participating in social and political movements in innovative ways, making use of numerous communication tools and reaching audiences worldwide.

The aim of this chapter is to explore how young people are shaping economics, society and global politics concerning the environment. With the growth of social media, the diffusion of information and the massive youth movements against climate change have been expanding around the world. The role of youth has gained particular importance in global politics, as they are increasingly claiming for policies addressing the failure of traditional political approaches regarding climate change. Lastly, with Millennials and Generation Z internalizing the environmental impact in their consumption decisions, firms taking actions to reach environmental standards and governments implementing policies towards greener and more efficient production and consumption, huge transformations in the economic area are being influenced by the youth.

This chapter is organized as follows: Section 7.2 provides insights into the relationship between youth and society, describes the main events and analyzes how movements are changing and how they are affecting the whole society. Section 7.3 provides an analysis of how the youth are getting involved in politics in response to global policy failures and the changes they have achieved on the international stage. Finally, Sect. 7.4 focuses on the main economic changes in relation to the environment and discusses the role of youth in the transition towards a greener economy.

7.2 Social Considerations: Youth and Climate Change

The effects of climate change have been well documented and predicted by the scientific community, particularly the Intergovernmental Panel on Climate Change (IPCC). Some even suggest calling it a "climate crisis." The impacts of climate change range from heat waves, droughts and floods to loss of biodiversity. For instance, climate change would impact extreme poverty, driving between 3 and 122 million to poverty by 2030 (Hallegatte et al. 2016), and the number of people vulnerable to sea level rise would increase to at least 410 million by 2100 (Hooijer and Vernimmen 2021). Thus, climate change has become a cause of political conflict and social unrest.

Multiple theoretical models have appeared since the 1960s to describe and analyze collective action, social movements and protests. The paradigms that existed at the time, the Marxist and the structural-functionalist models, to interpret social conflict were not quite adequate to understand claims such as those centered around women's liberation or environmental protection. For half a century, scholars developed theories to explain various aspects and reasons for collective action. Mario Diani defines social movements as a distinct social process, wherein actors "are involved in conflictual relations with clearly identified opponents; are linked by dense informal networks; share a distinct collective identity." (Della Porta and Diani 2006, p. 20). These three characteristics are present in young climate movements as they promote social change and specifically identify targets for their actions; they coordinate initiatives and define strategies through negotiations with individuals and organizations involved and share a collective identity that goes beyond specific events and initiatives (Della Porta and Diani 2006).

The reasons for social movements to arise between young people are connected to climate change effects. Children and young adults will face a grim future due to climate change, ecological degradation, migration, deeper inequality and conflict (Clark et al. 2020). Eskenazi et al. (2020) warn that youth's security, physical and mental health will be impacted by climate change. The availability of scientific information has shaped the narrative of the youth, who blame powerful actors and older people for the climate crisis (Han and Ahn 2020). In fact, Greta Thunberg, one of the most recognizable faces among the youth advocating for climate action, asked politicians to listen to the scientific community and enact climate change policies (Han and Ahn 2020). Youth activists have criticized the failures and unkept promises of politicians, more so because of greenwashing narratives and practices of decision-makers and their vested interests in fossil fuels (Sultana 2022).

Injustices linked to climate change have spurred youth into action to mobilize against this reality. Thus, "[c]limate justice paradigms have emerged to address this failure [to tackle the climate crisis] and young activists have been prominent leaders of such movements…" (Ritchie 2020, p. 2). Climate protests have been organized around the world, and the most notable are the "climate strikes" led by Greta Thunberg. These took place on Fridays, thus named "Fridays for Future," and went on for more than a year. On March 15, 2019, more than 1 million people took to the streets (Boulianne et al. 2020). Young people went on strike to draw attention to the need for climate action and pressure governments and the media about it (Boulianne et al. 2020). The number of people ultimately reached 7.6 million in September 2019 (350.org 2019).

These strikes and other manifestations have emphasized that young people are willing to act for the environment and climate justice (Hilder and Collin 2022). "[S]ocial media innervates youth political participation, helping to supply the flow of ideas, information, and support youth need to engage." (Maher and Earl 2019, p. 8). Thunberg was able to have an extensive reach thanks to social media. Jung et al. (2020) noted that social media allows for direct communication in comparison to traditional media; what is more, the authors highlight that these platforms are useful for spreading a message effectively to many. Other factors influence young people alongside social media, which has been an invaluable tool. The community, parents, schooling and peers have an effect on political participation, though with varying degrees of incidence. Discussion of politics with family members is particularly successful in increasing the level of participation, but peers are more influential (Quintelier 2013).

For social movements, there can be multiple challenges in scaling, coordination and sustainability in time. Social media has been useful to bypass such obstacles and strengthen networks for convening power and organization. Thus far, activists and groups have been able to group in various locations for months or years. What is more, scientific information has backed up their claims and helped them integrate more individuals into the movement. However, momentum may fizzle and worldviews may clash and cause ruptures. That being said, if desired objectives are accomplished, the movement's existence may not be pressingly necessary, as conflictual relations with power actors and desire for social change are fundamental characteristics.

As stated above, climate change has a profound impact on the lives of young people who are particularly concerned about the climate crisis, as forecasts paint a bleak picture. Youth rarely are included in political conversations, yet it can be harmful to democracies to exclude their views, experiences and knowledge instead of finding them valuable (Andersson 2015). Youth are often excluded from discussions, processes and political structures, as these are mainly focused on adult interests and needs (Harris et al. 2010). Habashi (2017, p. 22) points out that "[y]outh might be disenchanted by the system and refuse to engage in the expected political process, while simultaneously engaging in alternative political environments." Moreover, adultism is often a barrier for young people. Due to this, they often engage in other ways, such as lifestyle choices like veganism, recycling, boycotting or protesting (Pickard 2019).

The marginalization of young people is often tied to dominant views of them as "developing" or "unfinished" citizens, incapable of making political decisions (Gordon 2007). They also lack economic and social privileges and adults give them limited independence to interact in social and political activities, as "[c]hildren are synonymous with young humans that lack experience, knowledge, and insight due to their age." (Habashi 2017, p. 28). In fact, there is a myth that children and young people are dependent on adults and others until they grow (Stone 2021), but it ignores the power that exists in the youth, especially since the 1960s. Young people have been central players in movements and protests, which Han and Ahn (2020) exemplify with May 68, Tiananmen Square and the Arab Spring. "Yet today, youth are often framed in the mass media as, at best, apathetic, disengaged, and removed from civic action." (Constanza-Chock 2012, p. 1). Earl et al. (2017) argued that young people not only face barriers in their engagement, as it is assumed they are not interested in politics but are also denied agency in their own political socialization, as most models assume a top-down communication from adults to youth.

Young environmental activism is sometimes considered different from mainstream environmental activism, because of intersecting oppressions: youth is a "typology" of marginalization that intersects with others (Bowman 2020). In other words, it "...is a response to historical and structural inequalities, intersecting forms of disadvantage and multiple crises." (Hilder and Collin 2022). Various perspectives, such as intersectionality and ecofeminism, environmental justice, postcolonialism and decolonialism, propose to consider how people are shaped by multiple factors and realities. For instance, aspects like race, nationality, place, class and gender influence exposure and vulnerability to climate change around the world (Ergas et al. 2021). That is why "[p]oor, rural, Indigenous women and children in Global South nations are most vulnerable to climate-related disasters..." (Ergas et al. 2021, p. 21). This activism varies around the globe due to factors such as main climate concerns, political context or media environment (Eide and Kunelius 2021). While some activism is location-specific, some movements also seek to platform claims from disadvantaged communities, recognizing the disproportionate effects of climate change. For instance, there is often an emphasis on the lived experiences of indigenous people around the world and how imperialism and neocolonialism has destroyed their ecosystems, spirituality, knowledge systems and more.

Even though the environmental movement has existed for years and all around the world, its latest and most recognizable expressions have taken place in the Global North. However, claims for climate change action are not circumscribed to urban, middle-class and white communities. Rafaely and Barnes (2020) point out that African activists have highlighted for a long time that the continent's voice is excluded from global climate justice activism and that, particularly, African women are discriminated against by the media. Though Black women are at the forefront of environmental movements, their contributions are often overshadowed (Brooks, 2008 cited in Hasan 2024), which creates a biased representation that both reinforces power imbalances and lacks the diversity of realities and perspectives that actually exist (Hasan 2024). The erasure of Ugandan activist, Vanessa Nakate, from the 50th World Economic Forum coverage by the Associated Press is an example of this.

The core of youth climate activism seems to be the desire for structural change and systemic change: rebuilding the world to address the causes of the current climate crisis (Bowman 2019). Though it is difficult to measure their impact, young people seem to be successful in their efforts. They are agents of change in climate issues and their actions have created a sense of urgency in involving them in climate governance and policymaking (Han and Ahn 2020). As mentioned, the environmental movement has existed for several decades and youth's participation in it is not new; however, it is hard to deny that the impact recent initiatives and protests have had on public discourse. Young people have decided to organize in social movements to achieve their goals, even in the face of discrimination and marginalization. Though historically ignored in Western political discourse as mentioned, youth's involvement in climate change issues shows their power and potential (Ritchie 2020).

7.3 Youth Climate Movement: The Impacts on Global Political Stages

A lot of facts guide us to a disappointing conclusion: political leaders have done little in their commitment to addressing the global climate crisis. Directing international cooperation on environmental issues is exceptionally demanding. The measures and policies implemented thus far have proven inadequate to address global warming; instead, global leaders have raised fossil fuels subsidies (Carrington 2023). Among the twenty countries comprising the G20, a stark contrast emerged. Developed countries have put forward an ambitious target for carbon emissions reduction. Nevertheless, developing regions voiced their opposition to the commitment to achieving a 60% reduction in carbon emissions by 2030 (Paramasivam and Abnett 2023). While complicated dialogue delays climate targets, the power of the youngest around the world has grown, with the introduction of political changes. Both in the Paris Agreement in 2015 and Madrid's Conference 25 of the Parties (COP) in 2019, the youth assumed the central role in the climate justice movement (United Nations Climate Change 2019). They now have the chance to discuss how to act rapidly on climate issues, using the Paris Agreement as a blueprint. Youth movements have always had the power to introduce new cultural, political and educational changes. For example, in 1968, the United Nations Educational, Scientific and Cultural Organization (UNESCO) called on the head of state United Nations members to attend the first Intergovernmental Conference for the Rational Use and Conservation of the Biosphere. The main goal of this meeting was to discuss the challenging balance between development and the environment, by presenting scientific facts about impacts on nature (UNESCO 1968). The youth environmental movement, through protests and strikes, succeeded in getting world leaders to sit together and listen to what was happening with the environment. The World Program of Action for Youth, established in 1995, stated the importance of granting younger generations the right to participate at local, national, regional and international levels (United Nations 2010).

Between the 1990s and the first decade of the twenty-first century, youth political movements experienced a decline. In recent years, youth political mobilizations around climate change have been driven by several factors. Starting from new social theorists, one of the most remarkable characteristics of young movements is the sense of belonging to a group and the collective effort to fight for a common cause, like the climate crisis. Such movements are well known for their collective identities, which arise from individual identities and establish their boundaries based on ethnicity, region, social class, gender, religion and many other variables (Almeida 2020). The climate change crisis is a cause that strengthens the sense of unity among young activists, alongside other factors.

In addition, another important aspect driving down the engagement of young people in political movements is all the crises they have suffered since the 2000s (Henn et al. 2022). After the Wall Street financial crisis (2008–2009), youth employment and public spending suffered a great decrease, with austerity as a key player in most economies affecting in a negative way the youngest generation and their transit to adulthood. The waves of crises have led millennials and generation Z to be active in political protest asking for social, economic, labor and environmental policies and denouncing global leaders of doing nothing. Another fact that empowers the young political environmental activists is the unfairness of climate change, because the youth are the ones who most will suffer the impacts of global warming (Henn et al. 2022). What is more, O'Brien et al. (2018) argue that young social movements are based on the postmaterialist values, which have been brought about by wealthier economies and allow more people to access high levels of education. As a result, higher attendance to university has built a critical electorate who are not easy to manipulate and also are committed to human rights and the environment. In the same vein, Kolleck and Schuster (2022) state that the youth engaged in international climate negotiations are well-educated and that women are the ones who play a central role in youth international movements. In fact, having information about global issues made the newer generations stand up for change in front of political leaders in a different way that has not been done before, thanks to technology. In the twenty-first century, everything goes viral and everyone can be notified about what is happening in every corner of the world. Thereby, the younger generations use the media platforms like X (formerly Twitter), Instagram and TikTok to make the information go viral and to weave global networks.

Even though there is a worldwide youth environmental movement, which is making history, there are young people who do not share this worldview stated before. On the contrary, some studies have revealed that the young with lower income, which may lead them to less access to education, are the ones who support authoritarian-nationalist candidates such as Trump or Bolsonaro (Henn et al. 2022).

The United Nations Framework Convention on Climate Change (UNFCCC) is the forum where political global leaders negotiate, contest and agree on climate targets. As the youth have no structural power and tools, they hold a solid discursive attitude which gives them the chance to initiate normative changes at the international policy stage (Nisbett and Spaiser 2023). One way to make room for the youth, among the big global leaders, and to enter inside negotiation rounds is by making strikes and protests or by gaining agency in social media (Nisbett and Spaiser 2023). The impacts of youth advocacy networks across the world become evident when new norms such as anti-fossil fuels, phasing down unabated coal-fired power and phasing out fossil fuels subsidies, came up after a COP.

In 2016, in Marrakech, the youth movement had a success in the COP22. The group introduced the loss and damage fund as a central theme in climate change negotiations; by doing this they gained public attention and public relevance (Nisbett and Spaiser 2023). By 2018, in Katowice city, once again the youth made strides. They brought human rights back to the international agenda, advocating for the rights of women, children and indigenous communities. An example of the latter, thanks to the youth global network, an indigenous youth community from Northern Canada participated in the COP24. This youth community has raised their voices, expressing new ideas about climate governance and leadership (MacKay et al. 2020). The last main goal of the youth at the global stage was the new International Court of Justice resolution which establishes an advisory opinion on climate change and human rights. The core aim of these unanimous requests is to protect future generations from climate change impacts (World's Youth for Climate Justice 2023). The initiative surged from the Pacific Island Students Fighting Climate Change (PISFCC), joined by young people around the world and supported by the World's Youth For Climate Justice. At regional and national levels, for example in Europe, the Fridays For Future movement aimed to collect one million signatures, enough to pressure the EU parliament into prioritizing the climate crisis during 2018–2019 (Han and Ahn 2020).

In 2019, Germany declared that coal mining would be phased out by 2038 (Rinscheid and Wüstenhagen 2019). This green policy was achieved because thousands of young people in Germany joined different environmental movements. The young activists, through campaigns in person and on X (then-Twitter) and through strikes and manifestations, noted that Germany was not doing anything to protect the environment and to reduce CO2 emission. So, they pushed the Merkel administration to declare a climate policies package to phase out coal, reduce emissions, promote solar and wind energy and renovate houses to be more sustainable (Wehrmann and Wettengel 2021). It is important to note that in September 2021, before the national elections, the youth movement succeeded in setting up climate change policies, by carrying out national protests in hundreds of German cities. One of them was a hunger protest with the goal to meet the three leading chancellor candidates vying to replace Angela Merkel (Clean Energy Wire 2021). In North America, the Green New Deal, a national climate project, was achieved because of the young activists and the vocal advocacy of Congresswoman Alexandria Ocasio-Cortez, who promotes environmental policies (Tracy 2018). Additionally, the Sunrise Movement, led by youth, pushed for the Inflation Reduction Act during Biden's presidency, the largest environmental project in the USA, focusing on climate and clean energy investments to combat global warming (Sunrise Movement 2022). In the Global South, a movement called *Jóvenes por el Clima*, in Argentina, garnered national and regional recognition. In 2019, this group succeeded in the promotion of a resolution in the National Senate that declared the country is in a Climate and Ecological Emergency, the first country in Latin America and the fourth at global level; until now just Ireland, Canada and France were the only three countries to declare the climate emergency (La Nación 2019). Also, these young activists promoted a campaign for the approval of the Minimum Budget Law for Climate Change Adaptation and Mitigation (CLACSO 2020).

The examples stated above come from different regions, but they all demand action guided by scientific data and facts, like those provided annually by the IPCC. These large youth organizations around the world have access to technology that allows their movements to go viral. This enables young people to quickly organize and launch campaigns or strikes, which push political leaders to move forward towards green policies and climate change action. Politicians around the world cannot ignore these youth movements because they consistently attract support from various groups such as education, human and women's rights and labor justice movements. Furthermore, a pattern which has been identified is that the youth strategically organize strikes, campaigns or boycotts during key moments, such as election years, which are critical times for new candidates, as it happened in Germany.

Considering the facts discussed earlier, it becomes evident that youth play a key role across transnational, regional and national political arenas. Although they succeeded in presenting climate change as a pressing global issue of the highest priority, they lack the power to bring about immediate policy changes. They also lack representative legitimacy, so they face important limitations and challenges on the global stage. First, they are not able to alter the global system which is supported by those who maintain the status quo, consolidate their power and individual interests. These are the very actors who posses the power to control resources, public discourse and make room for different opinions. In addition, the leaders of countries with the highest CO_2 emissions dismissed the youth climate speeches and demands, such as Vladimir Putin, Donald Trump and Jair Bolsonaro (Han and Ahn 2020). In another note, there is one more challenge for the youth climate movement that stands out as particularly significant: the decline in youth civic engagement, as suggested by certain academics, appears to be closely linked to escalating social inequalities worldwide (O'Brien et al. 2018). This decline can be attributed to disparities in access to education, housing and employment, which ultimately diminishes the involvement of younger generations in civic activities. How can young generations engage in climate activism when they face challenges such as securing stable housing, establishing long-term job prospects and accessing higher education?

Certainly, these new generations have fought against lobbyists and climate change deniers with remarkable strength, highlighting that politicians are not acting urgently to address the climate crisis. The youth should be involved in political spaces at national, regional and international levels, as this will provide them with a lot of tools to develop the climate policies needed and also learn from actual politics, including the good decisions and the bad ones. They are the global future leaders that will have to deal with climate change impacts but with a worsened environment, if no action is really taken now. Finally, the intergenerational dialogue becomes indispensable as they are allies on global and local scales. With these partnerships between youth, non-governmental and governmental organizations, scientists, indigenous communities, women and civil society, the stage is set to introduce new ideas and climate blueprints that came from these networks, including the state heads.

7.4 Economics and the Environment: How the Youth is Shaping the Economy?

During the last decade, we have experienced a rapid increase in climate awareness at a global level, fueled by the acceleration of greenhouse gas emissions that started following the Industrial Revolution and continues until now. Since the Industrial Revolution, most socio-economic indicators, such as GDP, life expectancy, education and employment, have soared at unprecedented levels. Importantly, carbon emissions were not foreign to this process; rates at the global level rocketed in the eighteenth century following an impressive socio-economic transformation. In this context, concerns regarding carbon concentration leading to the acceleration and intensification of climate change have introduced a relevant question regarding the relationship between economic growth and pollution.

The benefits of economic growth and development are widely recognized and almost out of discussion today. Imposing limits on growth in order to reduce the accumulation of carbon dioxide would be costly, but does it mean that a climate crisis is inevitable? When studying the evolution of both economic growth and pollution for all countries, we do not always see a perfect positive relationship. Instead, for many countries and pollutants, studies find that this relationship follows an inverted U-shape, which has been referred to as the 'Environmental Kuznets Curve' in economics. This relationship between pollution and income indicates that at an initial stage, environmental deterioration will increase with economic growth, but as income continues growing this relationship inverts.

The first empirical work on the EKC was Grossman and Krueger (1991), in which they documented cross-sectional data for 42 countries and found this inverted U-shape relationship between pollution and per capita GDP focusing on sulfur dioxide and smoke. The hypothesis was further developed in the subsequent years (World Bank 1992; Grossman and Krueger 1995; Selden and Song 1994; Holtz-Eakin and Selden 1995; Stern et al. 1996; Stokey 1998) both through data collection and theoretical models.

However, this inverted U-shape relationship between pollution and per capita income does not hold for all countries (Azam and Khan 2016; Al Sayed and Sek 2013; Caviglia-Harris et al. 2009; Vincent 1997), or follows a slightly different pattern, and also studies that estimate the EKC find different turning points (i.e., the per capita GDP at which the positive relationship turns negative), depending on the data and methods, making it difficult to arrive to a general conclusion. Most estimates suggest a turning point below $6000 dollars; for example, Shafik (1994) and Torras and Boyce (1998) find a turning point at $4379 and $4641 1990 dollars (PPP), respectively, for sulfur concentrations using data from the Global Environment Monitoring System. However, some other estimations are far from that, such as Kaufmann et al. (1998), who found a turning point of $14,730 1990 dollars (PPP) for sulfur with data from the UN.

Explanations for pollution and growth following an inverted U-shaped curve include aspects on income distribution (Bimonte 2009), policy distortions (Panayotou 1997), political freedom (Torras and Boyce 1998), trade (Copeland and Taylor 1995; Suri and Chapman 1998; Tobey 1990; Antweiler et al. 2001; Reppelin-Hill 1999) and income elasticity of environmental quality demand (Pezzey 1992; Selden and Song 1994; Roca 2003; McConnell 1997; Shafik 1994).

Grossman and Krueger (1991) explained the observed relationship through scale, composition and technique effects. The scale effect captures the fact that the initial stage of growth requires intense use of resources that implies more emissions and deteriorates environmental quality; the composition effect refers to a structural transformation in the economy as income continue growing (Vukina et al. 1999; Panayotou 1997), while the technique effect contemplates the idea that income grow may lead to adoption of cleaner technologies (Komen et al. 1997) or regulations that conduct to pollution abatement. Moreover, many studies found regulation as the main factor explaining the observed relationship (Panayotou 1997; De Bruyn 1997). Explanations for increased regulation include societies assigning more importance to environmental quality once basic needs are met, higher-income countries allocating more budget for monitoring and higher income and education levels driving greater demand for environmental quality (Dasgupta et al. 1997; Pargal and Wheeler 1996; Selden and Song 1994).

The discussion of which factors explain environmental improvements in the context of the EKC presented above is interesting to think about the role of youths in the transition to a cleaner economy. The younger generations have a role in shaping emissions and growth through green consumption, demand for public policies, investment and adoption of green technologies, among others. In recent years, climate awareness has significantly increased worldwide, and younger people are generally more concerned than older people (Lowe et al. 1980). This awareness has a place in shopping decisions, as consumers are paying more attention to green consumerism, in part, due to rising environmental concerns (Reddy et al. 2023; Maturos Kanchanapibul et al. 2014). Manley et al. (2023) show that both Gen Z and Millennials desire to engage in sustainable consumption, although with small differences in their motives. Several studies have tried to identify factors influencing green consumption, for example, Gomes et al. (2023) show that environmental concerns, future green estimates and green perceived quality positively influence Generation Z's willingness to pay more for green products. Although many studies have shown that purchasing attitudes and actual buying behavior of consumers do not always match (Kollmuss and Agyeman 2002; Terlau and Hirsch 2015), others such as Brown et al. (2003) demonstrated that consumers who manifest intentions to buy products do actually purchase more than those who do not. This means that the increasing amount of people expressing intentions to buy more environmentally friendly products increases the probability of actual higher purchases of them.

The youth have a crucial role in the transition towards a greener economy not only via its purchasing decisions, but also because the growing emphasis on a green economy offers employment opportunities for them (Stanef-Puică et al. 2022). Most recent programs aiming to train and develop green skills to create more green jobs, understood as jobs that contribute to preserving or increasing environmental quality, are targeted at young people (Mwaura and Glover 2021).

From the supply side, there have been significant changes in the way firms behave regarding the environmental impact of their economic activities, which involves changes in the materials used in production, marketing campaigns highlighting their commitment to the environment, reporting of their impact and efforts to obtain distinctions of their products. Companies are increasingly making efforts to increase their environmental reputation, expected to add financial value to the company in the long run (Konar and Cohen 2001). According to a survey on green markets (European Commission 2022), most small and medium enterprises surveyed across the European Union are taking measures to be more resource efficient (89% are taking at least one action), and across them, the vast majority (83%) is planning to take more actions in the following two years.

Young people's attitudes towards the environment are intrinsically related to companies' behavior. Dragolea et al. (2023) stated that Generation Z is the real promoter of the Green Marketing culture, being the actual generation with responsible behavior regarding the environment. This makes green marketing practices much easier to transmit to Gen Z consumers and thus encourages companies to introduce them (Naidu et al. 2020). In this sense, social media is growing in relevance as a tool impacting consumers' decisions, particularly for the youth. Bryla et al. (2022) argued that social media has been progressing as a builder of both brand building and sustainable consumption, and Gong et al. (2023) demonstrated that there is a positive association between social network sites use and green consumption.

All those relevant transformations towards a greener economy have also a place in the stock market. The green bonds market has experienced rapid growth in recent years (Barua and Chiesa 2019), supporting a wide range of environmental-related projects. Agliargi and Agliardi (2019) identify environmental awareness as a key factor leading to higher demand for green bonds, which provide a cheaper way to fund environmentally related projects, while Li et al. (2019) suggest that the issuance of green bonds signal companies' responsibility towards the environment. As firms started disclosing information related to their targets and commitment to ESG (Environmental, Social and Governance), ESG ratings also became an important factor affecting the issuance of green bonds. Polbennikov et al. (2016) showed that there is a positive relationship between ESG scores and returns of bonds, while Wang and Wang (2022) show that Chinese listed firms with ESG practices can issue more green bonds. In the same direction, most countries have developed carbon markets where firms can trade carbon credits, including both the formal carbon market, which contemplates the EU carbon trading scheme and was designed in the context of the Kyoto Protocol, and the voluntary carbon market, which reflect the initiative of firms and organizations to actively find a solution to compensate for emissions that could not be reduced.

In the economic context, the youth have been a crucial part influencing significant changes towards a cleaner economy through their growing environmental awareness that affect consumption decisions, companies' behavior, environmental policies and green employment dynamics, among others. Remarkably, all these changes do not necessarily imply slowing economic growth, as new consumer demands are accompanied by the development of cleaner products and services, abatement technologies that make production less harmful to the environment and more informed decisions. That said, although the high rise in carbon emissions that followed the Industrial Revolution has been accompanied by a significant transformation of the economy and the institutions, involving consumers, firms and governments; we cannot assure that all the transformations mentioned in this section are sufficient to reach the targets imposed by the Paris Agreement. That is something that governments in each country should consider, they should learn from other countries' experiences and try to accelerate transition by keeping the focus on the role of the younger generations to make that possible.

7.5 Conclusion

As climate awareness increases, the whole society is being transformed in an effort to mitigate the negative consequences of climate change and adapt to new standards. The youth are a key part of this transformation. They are influencing all parts of society towards change with conscious consumption decisions, widespread information and activism.

In contrast with older generations, the youth are paying more attention to the environmental impact of their choices, therefore increasing the demand for products and services that are more environmentally friendly, creating positive spillover effects for firms to commit to environmental objectives and for governments to adopt policies that satisfy the demands for better environmental quality. Moreover, the global youth population has shown a growing interest in climate activism and advocacy, largely fueled by their disappointment with the world's political leaders. The inadequate management of leaders in tackling the climate crisis has sparked a remarkable upswing in youth-led global social movements. Not less important, social media has played a crucial role in the diffusion of information related to climate change. The youth has been exploiting the available communication tools to reach audiences and create networks at a global level, reshaping societal dynamics.

Overall, the youth are demonstrating commitment to the climate crisis and are implementing actions to mitigate the effects. We cannot say if all these efforts are sufficient to meet international climate objectives, but surely the youth are showing they can lead the needed transformations.

References

350.org (2019) 7.6 million people demand action after week of climate strikes. [Online] Available at: https://350.org/7-million-people-demand-action-after-week-of-climate-strikes/. Accessed 23 Oct 2023

Agliardi E, Agliardi R (2019) Financing environmentally-sustainable projects with green bonds. Environ Dev Econ 24(6):608–623. https://doi.org/10.1017/S1355770X19000020

Almeida P (2020) Movimientos sociales: La estructura de la acción colectiva (L. Mosconi, Trans.; 1st ed.). CLACSO. https://www.clacso.org.ar/biblioteca_pdf/almeida_movimientos.pdf

Alsayed A, Sek S (2013) Environmental Kuznets curve: evidences from developed and developing economies. Appl Math Sci 7:1081–1092. https://doi.org/10.12988/ams.2013.13098

Andersson E (2015) Situational political socialization: a normative approach to young people's adoption and acquisition of political preferences and skills. J Youth Studies 18(8):1–17. https://doi.org/10.1080/13676261.2015.1020926. Reprinted by permission of the publisher Taylor and Francis Ltd, http://www.tandfonline.com, license Number 5658430291794

Antweiler W, Copeland BR, Taylor MS (2001) Is free trade good for the environment? Am Econ Rev 91(4):877–908. https://doi.org/10.1257/aer.91.4.877

Azam M, Khan AQ (2016) Urbanization and environmental degradation: evidence from four SAARC Countries—Bangladesh, India, Pakistan, and Sri Lanka. Environ Prog Sustainable Energy 35:823–832. https://doi.org/10.1002/ep.12282

Barua S, Chiesa M (2019) Sustainable financing practices through green bonds: what affects the funding size? Bus Strategy Environ 28(6):1131–1147

Bimonte S (2009) Growth and environmental quality: testing the double convergence hypothesis. Ecol Econ 68(8–9):2406–2411. https://doi.org/10.1016/j.ecolecon.2009.03.020

Boulianne S, Lalancette M, Ilkiw D (2020) School strike 4 climate: social media and the international youth protest on climate change. Media Commun 8(2):208–218. https://doi.org/10.17645/mac.v8i2.2768. License: Creative Commons Attribution CC BY 4.0. https://creativecommons.org/licenses/by/4.0/

Bowman B (2019) Imagining future worlds alongside young climate activists: a new framework for research. Fennia 197(2):295–305. https://doi.org/10.11143/fennia.85151. License: Creative Commons Attribution CC BY 4.0. https://creativecommons.org/licenses/by/4.0/

Bowman B (2020) They don't quite understand the importance of what we're doing today: the young people's climate strikes as subaltern activism. Sustain Earth 3(16):1–13. https://doi.org/10.1186/s42055-020-00038-x. License: Creative Commons Attribution CC BY 4.0. https://creativecommons.org/licenses/by/4.0/

Brown M, Pope N, Voges K (2003) Buying or browsing? An exploration of shopping orientations and online purchase intention. Eur J Mark 37:1666–1684. https://doi.org/10.1108/03090560310495401

Bryła P, Chatterjee S, Ciabiada-Bryła B (2022) The impact of social media marketing on consumer engagement in sustainable consumption: a systematic literature review. Int J Environ Res Public Health 19(24):16637. https://doi.org/10.3390/ijerph192416637

Carrington D (2023) Fossil fuels being subsidized at rate of $13m a minute, says IMF. The Guardian, August 24. Available at https://www.theguardian.com/environment/2023/aug/24/fossil-fuel-subsidies-imf-report-climate-crisis-oil-gas-coal#:~:text=The%20IMF%20analysis%20found%20petrol,Russia%2C%20the%20EU%20and%20India. Accessed 24 Aug 2023

Caviglia-Harris J, Chambers D, Kahn J (2009) Taking the "U" out of Kuznets: a comprehensive analysis of the EKC and environmental degradation. Ecol Econ 68(4):1149–1159. https://doi.org/10.1016/j.ecolecon.2008.08.006

CLACSO (2020) Jóvenes por el clima. CLACSO. https://biblioteca-repositorio.clacso.edu.ar/bitstream/CLACSO/250282/1/Jovenes-por-el-clima.pdf

Clark H, Coll-Seck AM, Banerjee A, Peterson S, Dalglish SL, Ameratunga S, Costello A et al (2020) A future for the world's children? A WHO-UNICEF-Lancet commission. The Lancet 395(10224):605–658. https://doi.org/10.1016/S0140-6736(19)32540-1. License number: 5658431015677

Clean Energy Wire (2021, September 21) Hunger strikers demand meeting with chancellor candidates about climate. Clean Energy Wire. https://www.cleanenergywire.org/news/hunger-strikers-demand-meeting-chancellor-candidates-about-climate

Copeland BR, Taylor MS (1995) Trade and transboundary pollution. Am Econ Rev 85(4):716–737. http://www.jstor.org/stable/2118228

Costanza-Chock S (2012) Youth and social movements: key lessons for allies. Berkman Center Research Publication No. 2013-13. https://cyber.harvard.edu/node/97244. License: Creative Commons Attribution CC BY 3.0. https://creativecommons.org/licenses/by/3.0/

Dasgupta, S., Hua, W. and Wheeler, D. (1997) Surviving Success: Policy Reform and the Future of Industrial Pollution in China. The World Bank, 1. http://econ.worldbank.org/external/default/main?pagePK=64165259&piPK=64165421&theSitePK=469372&menuPK=64216926&entityID=000009265_3971229181112

de Bruyn SM (1997) Explaining the environmental Kuznets curve: structural change and international agreements in reducing sulphur emissions. Environ Dev Econ 2(4):485–503. http://www.jstor.org/stable/44379190

Della Porta D, Diani M (2006) The study of social movements: recurring questions, (partially) changing answers. In: Della Porta D, Diani M (eds) Social movements: an introduction. Blackwell Publishing, Cornwall, pp 1–32

Dragolea L-L, Butnaru GI, Kot S, Zamfir CG, Nuţă A-C, Nuţă F-M, Cristea DS, Ştefănică M (2023) Determining factors in shaping the sustainable behavior of the generation Z consumer. Front Environ Sci 11:1096183. https://doi.org/10.3389/fenvs.2023.1096183

Earl J, Maher TV, Elliott T (2017) Youth, activism, and social movements. Sociol Compass 11(e12465):1–14. https://doi.org/10.1111/soc4.12465.STMSignatoryPublishers

Eide E, Kunelius R (2021) Voices of a generation the communicative power of youth activism. Clim Change 169(9):1–20. https://doi.org/10.1007/s10584-021-03211-z. License: Creative Commons Attribution CC BY 4.0. https://creativecommons.org/licenses/by/4.0/

Ergas C, McKinney L, Bell SE (2021) Intersectionality and the environment. In: Schaefer Caniglia B et al (eds) Handbook of environmental sociology. Springer, Cham, Switzerland, pp 15–34

Eskenazi B, Etzel RA, Sripada K, Cairns MR, Hertz-Picciotto I, Kordas K, Machado Torres JP, Mielke HW, Oulhote Y, Quirós-Alcalá L, Suárez-López JR, Zlatnik MG (2020) The international society for children's health and the environment commits to reduce its carbon footprint to safeguard children's health. Environ Health Perspect 128(1):1–3. https://doi.org/10.1289/EHP6578

European Commission (2022) Flash Eurobarometer 498. SMEs, green markets and resource efficiency. Available at https://europa.eu/eurobarometer/surveys/detail/2287. Accessed 22 Oct 2023. License: Creative Commons Attribution CC BY 4.0. https://creativecommons.org/licenses/by/4.0/

Gomes S, Lopes JM, Nogueira S (2023) Willingness to pay more for green products: a critical challenge for Gen Z. J Clean Prod 390:136092. License Number: 5657251025157

Gong Y, Chen C, Tan Y, Tang D (2023) How active social network site use affects green consumption: A moderated mediation model. Front Psychol 14. ISSN: 1664-1078. https://doi.org/10.3389/fpsyg.2023.1124025. License: Creative Commons Attribution CC BY 4.0. https://creativecommons.org/licenses/by/4.0/

Gordon HR (2007) Allies within and without: how adolescent activists conceptualize ageism and navigate adult power in youth social movements. J Contemp Ethnogr 36(6):631–668. https://doi.org/10.1177/0891241606293608.STMSignatoryPublishers

Grossman G, Krueger A (1991) Environmental impacts of a North American free trade agreement. National Bureau of Economics Research, Working Paper, No. 3194. NBER, Cambridge. https://doi.org/10.3386/w3914

Grossman GM, Krueger AB (1995) Economic growth and the environment. Q J Econ 110(2):353–377. https://doi.org/10.2307/2118443. License Number: 5658901315589

Habashi J (2017) Reconceptualizing youth political socialization: a theoretical framework. In: Political socialization of youth. Palgrave Macmillan, New York

Hallegatte S, Bangalore M, Bonzanigo L, Fay M, Kane T, Narloch U, Rozenberg J, Treguer D, Vogt-Schilb A (2016) Shock waves: managing the impacts of climate change on poverty. Climate change and development series. World Bank, Washington, DC. https://doi.org/10.1596/978-1-4648-0673-5. License: Creative Commons Attribution CC BY 3.0 IGO. https://creativecommons.org/licenses/by/3.0/igo/deed.es

Han H, Ahn SW (2020) Youth mobilization to stop global climate change: narratives and impact. Sustainability 12(10):1–23. https://doi.org/10.3390/su12104127. License: Creative Commons Attribution CC BY 4.0. https://creativecommons.org/licenses/by/4.0/

Harris A, Wyn J, Younes S (2010) Beyond apathetic or activist youth: 'Ordinary' young people and contemporary forms of participation. Young 18(1):9–32. https://doi.org/10.1177/110330880901800103.STMSignatoryPublishers

Hasan M (2024) Symbolic annihilation: analyzing associated press's erasure of Vanessa Nakate and marginalization of black women activists. J Black Stud 55(3):253–267. https://doi.org/10.1177/00219347231220679

Henn M, Sloam J, Nunes A (2022) Young cosmopolitans and environmental politics: how post-materialist values inform and shape youth engagement in environmental politics. J Youth Stud 25(6):709–729. https://doi.org/10.1080/13676261.2021.1994131

Hilder C, Collin P (2022) The role of youth-led activist organisations for contemporary climate activism: the case of the Australian youth climate coalition. J Youth Stud 25(6):793–811, https://doi.org/10.1080/13676261.2022.2054691. License: Creative Commons Attribution CC BY 4.0. https://creativecommons.org/licenses/by/4.0/

Holtz-Eakin D, Selden T (1995) Stoking the fires? CO_2 emissions and economic growth. J Public Econ 57(1):85–101. https://doi.org/10.1016/0047-2727(94)01449-X
Hooijer A, Vernimmen R (2021) Global LiDAR land elevation data reveal greatest sea-level rise vulnerability in the tropics. Nat Commun 12(3592):1–7. https://doi.org/10.1038/s41467-021-23810-9. License: Creative Commons Attribution CC BY 4.0. https://creativecommons.org/licenses/by/4.0/
Jung J, Petkanic P, Nan D, Kim JH (2020) When a girl awakened the world: a user and social message analysis of Greta Thunberg. Sustainability 12(7):1–17. https://doi.org/10.3390/su12072707. License: Creative Commons Attribution CC BY 4.0 https://creativecommons.org/licenses/by/4.0/
Kanchanapibul M, Lacka E, Wang X, Chan HK (2014) An empirical investigation of green purchase behaviour among the young generation. J Clean Prod 66:528–536. https://doi.org/10.1016/j.jclepro.2013.10.062
Kaufmann R, Davidsdottir B, Garnham S, Pauly P (1998) The determinants of atmospheric SO_2 concentrations: reconsidering the environmental Kuznets curve. Ecol Econ 25(2):209–220. https://doi.org/10.1016/S0921-8009(97)00181-X
Kolleck N, Schuster J (2022) Youth participation in global policy networks on climate change. Int J Educ Res 114:102002. https://doi.org/10.1016/j.ijer.2022.102002
Kollmuss A, Agyeman J (2002) Mind the gap: why do people act environmentally and what are the barriers to pro-environmental behavior? Environ Educ Res 8(3):239–260. https://doi.org/10.1080/13504620220145401
Komen MHC, Gerking S, Folmer H (1997) Income and environmental R&D: empirical evidence from OECD countries. Environ Dev Econ 2(4):505–515. http://www.jstor.org/stable/44379191
Konar S, Cohen MA (2001) Does the market value environmental performance? Rev Econ Statis 83(2):281–289. http://www.jstor.org/stable/3211606
La Nación (2019, July 17) Argentina declaró la emergencia climática. La Nación
Li X, Du J, Long H (2019) Dynamic analysis of international green behavior from the perspective of the mapping knowledge domain. Environ Sci Pollut Res Int 26(6):6087–6098. https://doi.org/10.1007/s11356-018-04090-1
Lowe GD, Pinhey TK, Grimes MD (1980) Public support for environmental protection: new evidence from national surveys. Pac Sociol Rev 23(4):423–445. https://doi.org/10.2307/1388734
MacKay M, Parlee B, Karsgaard C (2020) Youth engagement in climate change action: case study on indigenous youth at COP24. Sustainability 12(16):6299. https://doi.org/10.3390/su12166299
Maher TV, Earl J (2019) Barrier or booster? Digital media, social networks, and youth micromobilization. Sociol Perspect 62(6):865–883. https://doi.org/10.1177/0731121419867697
Manley A, Seock Y-K, Shin J (2023) Exploring the perceptions and motivations of Gen Z and Millennials toward sustainable clothing. Fam Consum Sci Res J 51:313–327. https://doi.org/10.1111/fcsr.1247515523934
McConnell K (1997) Income and the demand for environmental quality. Environ Dev Econ 2(4):383–399. https://doi.org/10.1017/S1355770X9700020X
Mwaura G, Glover D (2021) Green jobs for young people in Africa: work in progress. Include evidence synthesis paper, July 2021. License: Creative Commons Attribution CC BY 4.0. https://creativecommons.org/licenses/by/4.0/
Naidu MN, Sonawane MS, Mohod MA, Pendam MB (2020) A study of consumers attitude towards green marketing. JournalNX—A Multidiscip Peer Rev J 65:83–91. License: Creative Commons Attribution CC BY 4.0. https://creativecommons.org/licenses/by/4.0/
Nisbett N, Spaiser V (2023) Moral power of youth activists—transforming international climate politics? Glob Environ Change (82). https://doi.org/10.1016/j.gloenvcha.2023.102717
O'Brien K, Selboe E, Hayward BM (2018) Exploring youth activism on climate change: dutiful, disruptive, and dangerous dissent. Ecol Soc 23(3). https://www.jstor.org/stable/26799169
Panayotou T (1997) Demystifying the environmental Kuznets curve: turning a black box into a policy tool. Environ Dev Econ 2(4):465–484. http://www.jstor.org/stable/44379189

Paramasivam P, Abnett K (2023) G20 nations fail to agree on emission reduction targets at talks. Available at https://www.reuters.com/sustainability/no-g20-consensus-emissions-mitigation-targets-talks-source-2023-07-28/. Accessed 28 July 2023
Pargal S, Wheeler D (1996) Informal regulation of industrial pollution in developing countries: evidence from Indonesia. J Polit Econ 104:1314–1327. https://doi.org/10.1086/262061
Pezzey JCV (1992) Economic analysis of sustainable growth and sustainable development. World Bank, Washington DC
Pickard S (2019) Young environmental activists are doing it themselves. Polit Insight 10(4):4–7. https://doi.org/10.1177/2041905819891364.STMSignatoryPublishers
Polbennikov S, Desclee A, Dynkin L, Maitra A (2016) ESG ratings and performance of corporate bonds. J Fixed Income 26:21–41. https://doi.org/10.3905/jfi.2016.26.1.021
Quintelier E (2013) Engaging adolescents in politics: the longitudinal effect of political socialization agents. Youth Soc:1–19. https://doi.org/10.1177/0044118X13507295
Rafaely D, Barnes B (2020) African climate activism, media and the denial of racism: the tacit silencing of Vanessa Nakate. Commun Psychol Glob Perspec 6(2/2):71–86. https://doi.org/10.1285/i24212113v6i2-2p71
Reddy KP, Chandu V, Srilakshmi S, Thagaram E, Sahyaja C, Osei B (2023) Consumers perception on green marketing towards eco-friendly fast moving consumer goods. Int J Eng Bus Manage. https://doi.org/10.1177/18479790231170962
Reppelin-Hill V (1999) Trade and environment: an empirical analysis of the technology effect in the steel industry. J Environ Econ Manag 38(3):283–301. https://doi.org/10.1006/jeem.1999.1085
Rinscheid A, Wüstenhagen R (2019) Germany's decision to phase out coal by 2038 lags behind citizens' timing preferences. Nat Energy 4:856–863. https://doi.org/10.1038/s41560-019-0460-9
Ritchie J (2020) Movement from the margins to global recognition: climate change activism by young people and in particular indigenous youth. Int Stud Sociol Educ 30(2):1–20. https://doi.org/10.1080/09620214.2020.1854830. Reprinted by permission of the publisher (Taylor and Francis Ltd, http://www.tandfonline.com), license number 5658430556199
Roca J (2003) Do individual preferences explain the environmental Kuznets curve? Ecol Econ 45(1):3–10. https://doi.org/10.1016/S0921-8009(02)00263-X
Selden TM, Song DS (1994) Environmental quality and development: is there a Kuznets curve for air pollution emissions? J Environ Econ Manag 27:147–162. https://doi.org/10.1006/jeem.1994.1031
Shafik N (1994) Economic development and environmental quality: an econometric analysis. Oxford Econ Papers 46:757–773. http://www.jstor.org/stable/2663498
Stanef-Puică MR, Badea L, Şerban-Oprescu GL, Şerban-Oprescu AT, Frâncu LG, Creţu A (2022) Green jobs-a literature review. Int J Environ Res Public Health 19(13):7998. https://doi.org/10.3390/ijerph19137998
Stern D, Common S, Barbier E (1996) Economic growth and environmental degradation: the environmental Kuznets curve and sustainable development. World Dev 24(7):1151–1160. https://doi.org/10.1016/0305-750X(96)00032-0
Stokey NL (1998) Are there limits to growth? Int Econ Rev 39(1):1–31. https://doi.org/10.2307/2527228
Stone L (2021) Youth power—youth movements: myth, activism, and democracy. Ethics Educ 16(2):249–261. https://doi.org/10.1080/17449642.2021.1896641. Reprinted by permission of the publisher (Taylor and Francis Ltd., http://www.tandfonline.com), license number 5652741092983
Sultana F (2022) The unbearable heaviness of climate coloniality. Polit Geogr 99(102638):1–14. https://doi.org/10.1016/j.polgeo.2022.102638.Licensenumber:5658470299649
Sunrise Movement (2022) Inflation reduction act. https://www.sunrisemovement.org/movement-updates/inflation-reduction-act/
Suri V, Chapman D (1998) Economic growth, trade and energy: implications for the environmental Kuznets curve. Ecol Econ 25(2):195–208. https://doi.org/10.1016/S0921-8009(97)00180-8

Terlau W, Hirsch D (2015) Sustainable consumption and the attitude-behaviour-gap phenomenon—causes and measurements towards a sustainable development. Int J Food Syst Dyn Int Center Manage Commun Res 6(3):1–16

The World Bank (1992) World development report 1992. Development and the Environment. https://doi.org/10.1596/0-1952-0876-5

Tobey J (1990) The effects of domestic environmental policies on patterns of world trade: an empirical test. Kyklos 43:191–209. https://doi.org/10.1111/j.1467-6435.1990.tb00207.x

Torras M, Boyce J (1998) Income, inequality, and pollution: a reassessment of the environmental Kuznets curve. Ecol Econ 25:147–160. https://doi.org/10.1016/S0921-8009(97)00177-8

Tracy M (2018) Alexandria Ocasio-Cortez's historic win and the future of the democratic party. The New Yorker. https://www.newyorker.com/magazine/2018/07/23/alexandria-ocasio-cortezs-historic-win-and-the-future-of-the-democratic-party

UNESCO (1968) Intergovernmental conference of experts on the scientific basis for rational use and conservation of the resources of the biosphere. Available at https://unesdoc.unesco.org/ark:/48223/pf0000067785/PDF/067785engo.pdf.multi. Accessed on 30 Oct 2023

United Nations (2010) World programme of action for youth. Available at https://www.un.org/esa/socdev/unyin/documents/wpay2010.pdf. Accessed on 30 Oct 2023

United Nations Climate Change (2019) Youth urge climate action at COP25. Available at https://unfccc.int/news/youth-urge-climate-action-at-cop25. Accessed on 30 Oct 2023

Vincent JR (1997) Testing for environmental Kuznets curves within a developing country. Environ Dev Econ 2(4):417–431. http://www.jstor.org/stable/44379186

Vukina T, Beghin JC, Solakoglu EG, Solakoglu EC (1999) Transition to markets and the environment: effects of the change in the composition of manufacturing output. Environ Dev Econ 4(4):582–598. http://www.jstor.org/stable/44404286

Wang S, Wang D (2022) Exploring the relationship between ESG performance and green bond issuance. Front Public Health. https://doi.org/10.3389/fpubh.2022.897577

Wehrmann B, Wettengel J (2021) Tracking climate and energy in Germany's election year. Elections and Politics. https://www.cleanenergywire.org/news/road-btw21-tracking-climate-and-energy-germanys-election-year

World's Youth for Climate Justice (2023) About the ICJ Advisory Opinion on Climate Change and Human Rights. Available at: https://www.wy4cj.org/icjao-proceedings (Accessed on 15th September 2025)

MIX
Papier aus verantwortungsvollen Quellen
Paper from responsible sources
FSC® C105338
www.fsc.org

If you have any concerns about our products,
you can contact us on
ProductSafety@springernature.com

In case Publisher is established outside the EU,
the EU authorized representative is:
Springer Nature Customer Service Center GmbH
Europaplatz 3, 69115 Heidelberg, Germany

Printed by Libri Plureos GmbH
in Hamburg, Germany